Abraham Malley

Micro-photography

Including a description of the wet collodion and gelatino-bromide processes

Abraham Malley

Micro-photography
Including a description of the wet collodion and gelatino-bromide processes

ISBN/EAN: 9783337142056

Printed in Europe, USA, Canada, Australia, Japan

Cover: Foto ©berggeist007 / pixelio.de

More available books at **www.hansebooks.com**

MICRO-PHOTOGRAPHY.

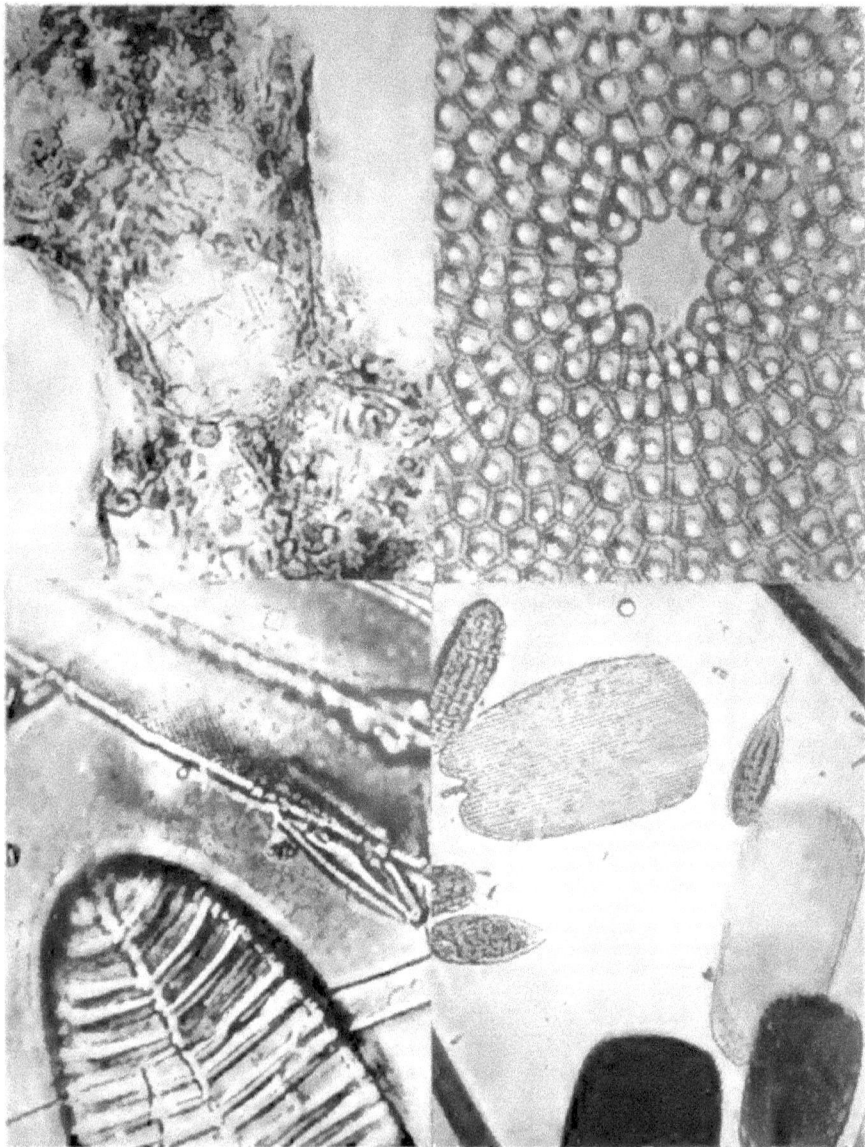

1. SECTION OF LUNG, CONTAINING BACILLUS ANTHRACIS FROM A CASE OF SPLENIC FEVER. 420 diameters.

2. CENTRAL PORTION OF A. KITTONII. 1500 diameters.

3. S. GEMMA AND P. SPENCERII. 1000 diameters.

4. SCALES OF P. ARGUS. 320 diameters.

MICRO-PHOTOGRAPHY

INCLUDING A

DESCRIPTION OF THE WET COLLODION

AND

GELATINO-BROMIDE PROCESSES

*TOGETHER WITH THE BEST METHODS OF MOUNTING
AND PREPARING MICROSCOPIC OBJECTS
FOR MICRO-PHOTOGRAPHY*

BY

A. COWLEY MALLEY, B.A., M.B., B.Ch., T.C.D.

LONDON
H. K. LEWIS, 136 GOWER STREET, W.C.
1883

PREFACE.

THE publication of this little work has been undertaken with a view to encourage the practice of Micro-photography, and afford a ready means of obtaining information without the trouble of refering to the innumerable papers on the subject, scattered up and down several scientific periodicals and publications.

As far as possible we have mentioned the sources from which our information has been derived; many have no doubt been overlooked, but the absence of a library for reference, and our isolated position, must serve as an excuse.

Our thanks are due to several friends, for their kindness in supplying any information required, but especially to the Rev. G. B. Powell who revised all the proofs, and whose valuable services have contributed in no small measure to hasten the completion and publication of the work.

A. COWLEY MALLEY.

Munslow, Shropshire.

CONTENTS.

CHAPTER I.

CHAPTER II.

CHAPTER III.

CHAPTER IV.

CHAPTER V.

CHAPTER VI.

CHAPTER VII.

CHAPTER VIII.

MICRO-PHOTOGRAPHY.

CHAPTER I.

INTRODUCTION.

BEFORE beginning a detailed description of the apparatus employed in Micro-photography, a few remarks are necessary on its utility in research, and its application to the various branches of science.

All microscopists are aware of the fickleness with which objects display their structure,—markings easily perceived at one time may baffle all attempts to resolve them at another, and perhaps when seen are beyond our power to delineate, I have before me at present some drawings which must have taken the artist a week to finish, and even then shew the existence in his mind of preconceived notions of their structure.

Photography not only obviates the necessity of future trouble and perhaps failure in the display of these markings, but in a period varying from a few seconds to as many minutes imprints the latent image on a sensitive surface, which after development, can be multiplied a thousand fold, and give to the world results, indisputably proving the truth of some favourite theory, or shewing the existence of some doubtful structure.

B

In the domain of pathology we find many observers differing in their descriptions of well-known lesions. What scientific man engaged in the investigation of the markings of the diatomaceæ agrees with others as to their true interpretations, at the present day Agnosticism is the only name expressing the condition of men's minds on some of these subjects. Thousands of questions hitherto unsettled on account of errors of description, or want of agreement between observers, are sure to find an easy solution if the practice of photography becomes more universal among naturalists and our professional brethren.

I do not wish to lead others to infer from this, that the photographic image may not be false if improperly obtained, but it can neither add anything to, nor take anything from, the structure, and may be relied on to shew what was actually seen by the observer.

In teaching Histology, Pathology, or any subject in which the microscope plays an important part, a photograph of the object may be thrown on the screen in the usual manner, and the lecture proceed uninterrupted by the use of separate apparatus, or the time taken up by those manipulations which must be made for every different observer.

It is unnecessary to dwell on the fascination of the pursuit, or the recreation afforded to those who after the arduous duties of the day, relieve the strain by the study of microscopy, for from what has been said the advantages of photography to the Histologist, Pathologist, and student of Natural History,

are so obvious that any tendency to extend its
domain to amusement might injure its prestige
with students of science.

In the following chapters our aim has been, not
so much to supply new methods as to increase the
popularity of those already known, by shewing the
facility of their application.

CHAPTER II.

PROPERTIES OF LENSES, IMPERFECTIONS OF RESULTING IMAGES, ABERRATIONS, ETC.

ACCORDING to the undulatory theory of light an object becomes visible, when it communicates the rapid vibratory motion of its molecules to the luminiferous ether, which being propagated through it in the form of spherical waves, is transmitted to the retina.

The vibration of the molecules takes place at right angles to the direction of the wave, for example, if a long string of beads is shaken at one end, the vibrations of the beads are at right angles to the length of the string, while the waves propagated through it by the shake move in the same direction as the length.

Reflection. When a luminous ray falls, or is incident, on a polished surface, it is reflected at an angle which equals its angle of incidence. (BAC and B″AC Fig. 1.)

Refraction. When a luminous ray passes from a rarer to a denser medium it is bent out of its course.

The Sines of the angles of incidence BAC, and refraction B′AC′, (Fig. 1), bear a constant ratio to each other.

This ratio is called the refractive index.

Refractive Index.

Of Air	1·000294.
„ Diamond	2·44
„ Flint Glass	1·642
„ Crown Glass	1·550

FIG. 1.

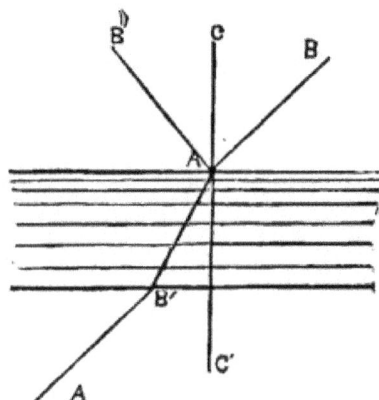

If the medium through which the light passes has parallel faces, the emergent B'A' and incident rays BA are parallel, (Fig. 1), if it has not, the tendency of the ray is to bend away from the angle at which the faces meet.

Lenses are of six forms; double convex 1, plano convex 2, converging meniscus 3, double concave 4, plano concave 5, diverging meniscus 6.

FIG. 2.

All rays falling on Convex lenses, are refracted towards the centres of curvature of the faces, those incident on Concave away from the centres, (Figs. 3 and 4).

In convex lenses the PRINCIPAL FOCUS (F Fig. 3) is that point where all rays parallel before entrance converge after refraction, it almost coincides with the centres of curvature of the faces.

If the rays diverge from a point A outside the principal focus, they meet after refraction on the other side beyond it B, these points A and B are *real foci*, (Fig. 3).

If they converge they meet at C within it, B and C are called conjugate foci.

Rays diverging from a point C Fig. 3, within the principal focus of a convex lens, can never meet at the opposite side, but only become less divergent, they therefore cannot form a real focus. But if the emergent rays are prolonged backwards they will

Fig. 3.

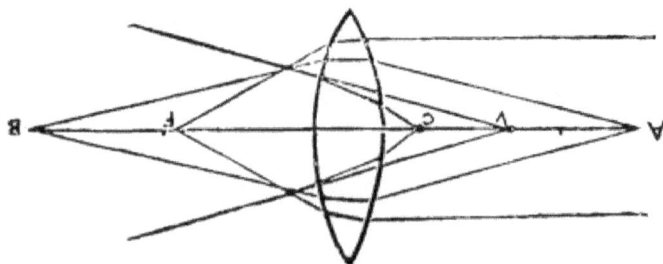

meet at V Fig. 3. The point V is called the *virtual focus.*

If an image of an object be received on a screen placed opposite a small hole in the shutter of a darkened room it will appear inverted, because the rays from different points of the object cross at the aperture.

FIG. 4.

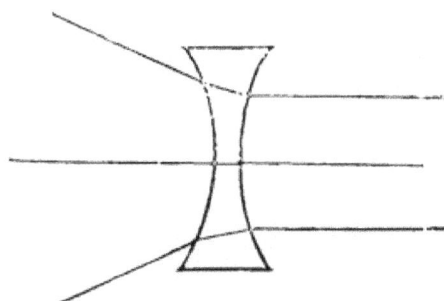

The size of the image varies inversely as the distance of the object, and directly as the distance of the screen.

If a convex lens be placed in the aperture, inverted and diminished images of distant objects will be formed on a screen placed at the principal focus.

If an object be placed at twice the focal distance of a convex lens, a real inverted image the same size as the object will be formed on a screen placed at a similar distance the opposite side.

FIG. 5.

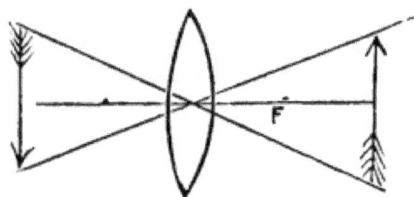

If nearer, a real inverted magnified image is formed at increasing distance until the principal focus is reached, when no image is formed on the screen, because the rays can never come to a focus being parallel.

FIG. 6.

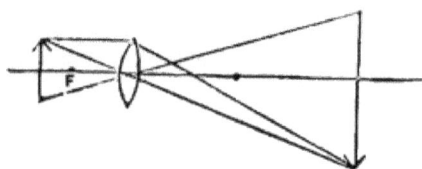

The human eye consists of a convex lens, the

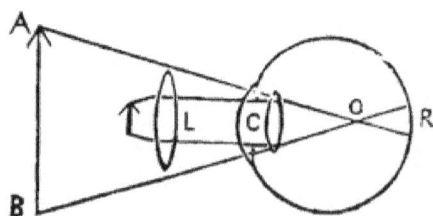

crystaline lens placed behind the aperture of the iris, and a screen, the retina, behind the crystaline lens, C and R.

A real inverted image of objects is formed on the retina, (how they appear erect to us has not yet been shown), and as the crystaline lens can alter its form, so as to bring convergent or divergent rays to a focus on the retina, the eye can perceive clearly the same object when placed at different distances.

There is a limit to the accommodation of the eye, in normal eyes eight or ten inches is the shortest distance an object can be clearly perceived without giving rise to an unpleasant feeling, owing to our efforts to increase the curvature of the lens, and enable it to bring the divergent rays to a focus on the retina, but if a convex lens L Fig. 7 be placed between the object and the eye it assists in bringing the rays to a focus on the retina.

The size of the retinal image varies with that of the angle AOB Fig. 7. This angle is called the visual angle.

As the image of an object is only a collection of the foci of the different points of an object, these images will be real or virtual as the image is situated without or within the principal focus.

A real magnified and inverted image of an object is seen by the eye if the object is situated outside the principal focus of a lens, and at less than twice the focal distance, because the angle of intersection of the rays from opposite sides of the object, e.g. the visual angle is increased.

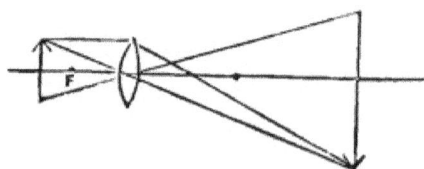

A virtual erect and magnified image is seen by the eye if the object is situated within the principal focus of the lens, because the crystaline lens of the observer C Fig. 7 has the power of bringing the rays to a focus and forming a real image on the retina R, on this principle the magnified image in the simple microscope depends.

Concave lenses can only formvirtual images no matter what the distance of theobject. (Fig. 4).

Up to the present the passage of light through lenses has been considered in its simplest form, it remains now to investigate the imperfections in the resulting images, their causes, and the various methods of correcting them.

When a ray of light passes from one substance through another a part is reflected at the incident surface, a second absorbed in its passage, and a third transmitted. The greater the thickness and number of the transparent media through which the light is transmitted, the greater is the quantity reflected and absorbed. Since Microscopic Objectives are composed of several lenses the loss of light is consequently very great.

Spherical Aberration shows itself in want of distinctness in an image at the margins when sharp at centre and *vice versa*, it is caused by the unequal refraction of the rays owing to the spherical shape of the lens, those at the margins coming to a focus before those at the centre.

Chromatic Aberration; a ray of white light is composed of rays of different colours, if it pass

FIG. 8.

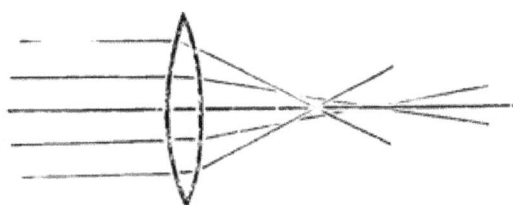

through a medium (such as a lens or prism), Fig. 8, whose sides are not parallel, it is divided into its component colours, or dispersed.

FIG. 9.

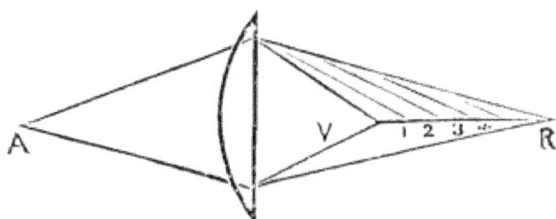

Now since all the different rays into which a pencil of white light can be divided, have different foci, (e.g. the point where the red rays come to a focus is more distant than that of the violet), the inter- mediate colours forming other foci V I 2 3 4 R Fig. 9 between them, are all unequally dispersed and the resulting image will be coloured at its edges.

There are several methods of remedying these aberrations, some of them diminish Spherical and Chromatic aberration at the same time.

Spherical Aberration may be destroyed by giving

the lens a decreasing curvature towards the edges, in fact an ellipsoidal form, the difficulty of this in practice was found so great that it was abandoned, and if the form was such that its performance was perfect for parallel rays, it would be inaccurate for all others.

It is much easier corrected by altering the curves of the two sides. For instance if the plane side of a plano-convex lens be turned towards a very near object, or the convex side towards a very distant one, the resulting aberration would be about one-fourth of that which would occur if the respective positions of the sides of the lens were reversed.

A method which has been frequently adopted for photographic objectives deserves mention, namely lessening the aperture, by placing a diaphragm in front of or behind the lens which admits the rays at the centre, but cuts off those at the edges, it will be evident that this lessens the chromatic aberration at the same time.

The diaphragm has other advantages, it enables objects lying at different distances from the lens to appear distinct, in other words, the penetrating power is increased. Its disadvantages are that it produces distortion, if placed in front of a lens the image of a square becomes barrell-shaped, if behind the reverse. It lessens the light, and diminishes the Angular Aperture, upon the extent of which the most important qualities of an objective depend.

The aberration of a concavo lens being exactly opposite to that of a convex lens, one may be made

FIG. 10.

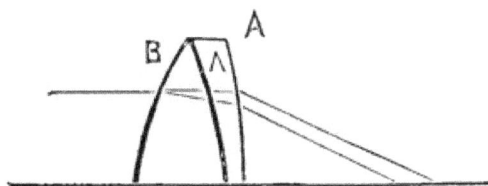

to correct the other without decreasing the magnifying power to any great extent. The lens A Fig. 10 completely destroys the spherical aberration of the lens B, for it does not change the position of the rays situated near its centre as much as it does those at the edges, owing to its increasing prismatic form.

If advantage be taken at the same time of the different relations between the refractive and dispersive powers of different kinds of glass, and if the lenses which correct the spherical aberration by their shape are formed from different kinds, chromatic aberration will be destroyed. If the lens A (Fig. 10) of high dispersive power, concave form, and low curvature, be joined to B of lower dispersive power and greater curvature, it is evident that the dispersion caused by the latter is neutralised, whilst its refractive power is only decreased by the opposite refraction of the concave.

The convex lens of crown is generally corrected by a concave of flint, it has been found that no two lenses can be made to correct each other perfectly, so opticians overcome the difficulty by combining several.

Before leaving this subject we should mention that in photography errors frequently arise, especially when using low powers, owing to the difference between the visual and chemical foci.

A point of white light may appear perfectly distinct at R Fig. 9, when surrounded by a violet areola, and if a photograph be taken it will be quite indistinct; while at V nearer the lens the image may not appear so distinct, but will be surrounded by a red areola, and if a photograph be taken at this point it will be found perfect, as the violet rays are the most chemically active.

This is owing to the different refrangibility of the red and violet rays, the latter coming to a focus sooner than the former; when photographing, the point V should be chosen when the object is surrounded by a red areola, this being the most accurate for the violet rays.

Microscopic objectives have of late years arrived at such perfection that it seems impossible to improve them.

They are designated from four inch up to one fiftieth of an inch, according to their magnifying power: for example, a one inch objective is supposed to magnify as much as a single lens of one inch focus, although its own focal length may be different.

This rule is only approximately correct, as glasses of the same designation by different makers vary in magnifying power.

Low powers up to two inches generally consist of a single combination, but as it has been found impracticable to construct objectives of high power free from chromatic and spherical aberrations in this way, all good makers have adopted the plan of correcting one combination by another, so that objectives from two inch to one inch generally consist of two, while those of quarter inch and upwards generally consist of three combinations, formed perhaps of as many as eight different lenses.

Various devices have been adopted to lessen the aberrations, increase the angular aperture, and at the same time reduce the number of lenses.

For an example of the ingenuity displayed in overcoming these difficulties, we refer the reader to the description of Mr. Wenham's new objective published in the *Proceedings of the Royal Society*, Vol. xxi, p. 111, and content ourselves with mentioning that it consists of a single front of the usual form, a single plano-convex back whose focus is four and one half times that of the front, and a middle triple in which a single concave lens of flint, three times the focal length of the front, corrects all the others.

In this combination only five lenses are used, and the errors arising from the sixteen surfaces of glass in the older forms reduced to ten.

1. A good objective should possess the Standard screw, adopted by the various societies, and now fixed by the best makers to their lenses.

2. Its definition should be clear, the field flat, objects at the edge of the field should be as free

from colour, and their definition as perfect as those
at the centre.   The illumination should be white, not
yellow, as is the case with inferior glasses, especially
those of foreign manufacture; if of low power, that
is of one inch or more nominal focal length, opaque
objects should be clearly shown without the use of
any apparatus for condensing the light upon them.

The rotundity of certain objects should come out
well, the more distant parts being in as good focus
as those nearer the objective.

Transparent objects possessing a certain thickness,
should be clearly defined to a certain depth.

This power of penetration is difficult to form a
correct opinion of, as we shall find that a glass of
given focus magnifies the depth of an object in the
proportion of the square of the magnification of the
diameter laterally.

It is in inverse ratio to the magnifying power of
the objective.

Therefore low powers are most suitable for the
display of those objects in which different points,
lying on different planes require to be in focus at the
same time.

3. It has been found that the thin glass used for
covering microscopic specimens, produces by refraction of the rays passing through it a sufficient negative aberration (Fig. 11) to destroy the adjustment
of an objective perfect in its performance on uncovered objects.

This may be remedied, as has been shown by Mr.
A. Ross, by under-correcting the front and over-

Fig. 11.

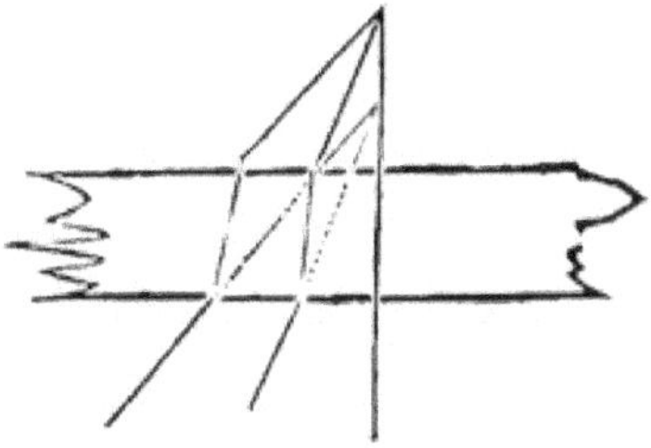

correcting the two back combinations, at the same time making the distance between them susceptible of alteration by means of a screw collar, this by bringing the front and back combinations nearer together, enables us to give the objective an excess of positive aberration sufficient to counteract the negative produced by the thin glass.

The screw collar adjustment at the same time corrects the errors caused by the difference in the refractive indices of the various media employed for the preservation and mounting of specimens. We therefore consider it an indispensable adjunct to all objectives of higher power than one half inch. It must not be forgotten that these adjustments affect the magnifying power considerably, and allowance must be made for this when accurate measurements of a covered object are required.

Many methods of practically making these corrections are mentioned in the various manuals, but I think the following will be found the easiest, and sufficiently correct for all purposes.

Turn the index (A Fig. 12) to Zero, in other words

Unity, then commencing with the lowest numbers this reaches as far as 1·52 for oil immersions.

This is called the Numerical aperture and NA is written after the numbers to distinguish it.

The angular aperture may be increased by using oblique light, which enables the rays illuminating the object to pass through at a greater angle, or an achromatic condenser, which increases both their number and obliquity.

We cannot leave this part of the subject without reverting to the views enunciated by Professor Abbe of Jena, which have placed the old theories of Microscopic vision on an entirely new footing.

According to his demonstrations the images produced by minute objects, small multiples of wave lengths, are imaged in an entirely different way from those produced by coarse objects.

Objects less than $\frac{1}{2500}$ of an inch are imaged by the diffracted rays produced by the action of the minute structure. The more diffracted rays reach the eye the truer the structure.

The more minute the object the more the diffracted rays are spread, therefore the larger the angular aperture the greater the quantity of diffracted rays transmitted and the truer the resulting image as regards the actual structure of the object.

It has been recently remarked that microscopic vision was totally distinct from macroscopic.

First, because by placing a stop behind the objective that is decreasing the aperture, objects could be made to assume different forms.

Second, because under different conditions of aperture different objects assumed the same form.

Now Professor Abbe himself says "minute structural details are not as a rule imaged by the *Microscope Dioptrically*, in accordance with the real nature of the object."

Some, less enamoured of the metaphysical aspect of the question, may find it difficult to perceive that images formed optically and dioptrically give us just as little information as to the real nature of the object.

If two scarlet cloths the same size, one marked with black squares the other with circles, and one coarser than the other, be placed at such a distance that their images subtend small angles, they will appear identical. If brought nearer, or what is the same thing, made to include a larger angle, a greater number of rays are collected, and certain differences appear, and as the angle of vision increases, the structural peculiarities still continue to cause fresh visual phenomena.

Professor Abbe goes on to say "and cannot be interpreted as Morphological but only as Physical characters, not *as images of material forms* but as signs of material differences in the nature of the particles composing the object, so that nothing more can safely be inferred from the image as presented to the eye, than the presence in the object of such structural peculiarities as will produce the particular diffraction phenomena, on which the images depend."

The wonderful discovery is that the Microscope

does not act dioptrically under certain circumstances, and that we have a new aid to research in these diffraction phenomena; like polarisation and other properties of light.

Unfortunately the whole stress has been laid upon the fact that these images are not exact representations of material forms.

Association and inherited materialism has taken too great a hold upon our minds, and we have forgotten that the same thing was proved with regard to all images many years ago; namely that nothing more can safely be inferred from the image presented to the eye, than the presence in the object of such structural peculiarities, as will produce the particular *refraction* phenomena on which the image depends.

Almost all scientists are misled by the error of conceiving material things to exist *apart* from the mind that perceives them, we do not mean by this that their *esse* is a mere *percipi*.

Some of our foreign brethren have justly said, " the English are that nation in Europe which like the Huckster and Workman Class of the state are destined to pass their lives immersed in matter." We raise a shout of astonishment at an old idea, no doubt in a new dress, and we miss the great discovery itself in our delight of at last perceiving by it, what has been dimly present to *our* minds for centuries.

These diffraction phenomena must not be confounded with those appearances surrounding the outlines of an object, mentioned in a previous paragraph when speaking of correction for cover glass.

For low powers and the majority of pathological investigations very large angular aperture is not considered advantageous, as with every increase of it the penetration and working distance decreases.

We hope that these explanations will enable the reader to choose an objective suitable for the particular branch of study he may be engaged in.

The extra amount expended in the purchase of first-class objectives, will produce a considerable saving in subsequent expenses, on account of the increase of successful results due to their employment. We therefore strongly recommend the amateur to patronise only the best makers, and pay the highest price for his objectives.

We shall devote a little time to the consideration of the principles involved in the formation of the image in the compound microscope, although we are not directly concerned with it in microphotography.

The compound microscope in its simplest form is composed of two lenses the objective and eye lens; fig. 13, 1 and 2, it will be perceived that the real

Fig. 13.

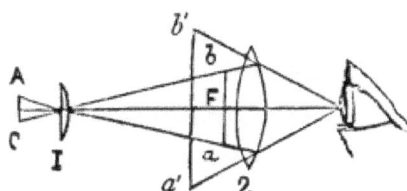

image a b of the object A B formed by the objective at a point just within the focus of the eye-glass is

changed into a virtual image at a' b' by the latter,
both images being inverted as regards the object.
The principle of the more complex forms of the compound microscope is similar, the number and arrangement of the lenses being different both for objective
and eye-piece, the former has been already explained.

The eye-piece generally used is the Huygeian
named after its inventor Huygens, it consists of two
plano convex lenses, 1 and 2 fig. 14, called respec-

Fig. 14.

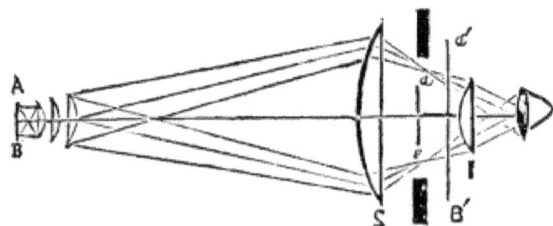

tively eye and field lens, the former being one-third
the focus of the latter and the distance of their
separation half the sum of their focal lengths.

Huygens was ignorant of one advantage this combination possessed, he adopted it simply to remedy
spherical aberration, but it has been since shown
that it corrects chromatic aberration also. A reference to the figure will explain the principle on which
the formation of the image depends.

We advise the amateur not to use it in photography except to view the image before focussing,
as with all its supposed advantages its employment
never gives good results.

CHAPTER III.

THE MICROSCOPE, ILLUMINATION, ETC.

Before beginning a detailed description of the microscope, a few hints to the uninitiated are necessary. Avoid purchasing instruments adorned with innumerable screws and mechanical arrangements for performing the various movements.

All superfluous apparatus, is not only useless, but a decided hindrance to accurate investigation, in fact the various additions generally made to the first class instruments of the present day, are an insult to the skilled microscopist and a means of perpetuating clumsy manipulations. On the other hand many excellent instruments are made, which, owing to some peculiarity of shape, are useless for photography.

The first essential of a microscope is a firm heavy stand, so arranged that the body may be inclined at any angle between horizontal and perpendicular and clamped in that position. It should be high enough to allow space beneath the stage for sub-stage apparatus, and the points on which it rests should be sufficiently distant from each other to prevent the instrument being easily overturned.

The coarse adjustment is best accomplished by rack and pinion movement, (not by one tube sliding within another) the rack being attached to a bar placed some distance from the tube carrying the eye-piece and objectives. When attached to the tube itself it is impossible to adopt any arrangement for shortening the tube, the consequence of this is that the surface of the ground glass in the camera, covered by the image, is greatly contracted owing to the length of tube between the objective and mouth of the bellows.

The fine adjustment should also be separate from the eye-piece tube to allow of attachments necessary for working it from a distance. It should respond immediately to the slightest rotation in either direction, and the object should suffer no apparent shake or displacement, when moving in or out of focus.

Nothing belonging to the microscope is susceptible of more improvement than the fine adjustment, and anything which would supersede the present very inferior methods by which this is effected, would be welcomed by microscopists.

The eye-piece tube must be removable from the body, or better from another tube about an inch long attached to the arm carrying the objectives. The way this is accomplished is of no importance as long as it is possible.

The stage should be firm, and as thin as is compatible with the use of the sub-stage apparatus, to be described. Circular in form and possessing some arrangement by which movements in a rectangular, circular and horizontal direction may be smoothly and easily communicated to the object, by aid of the finger alone without the use of milled heads or other mechanical arrangements.

This is best accomplished by ivory points moving on a blackened glass surface, a method extensively employed by some of our leading opticians and likely to supersede all others where real work is the object in view, not a desire to catch the eye of the public.

Fig. 15.

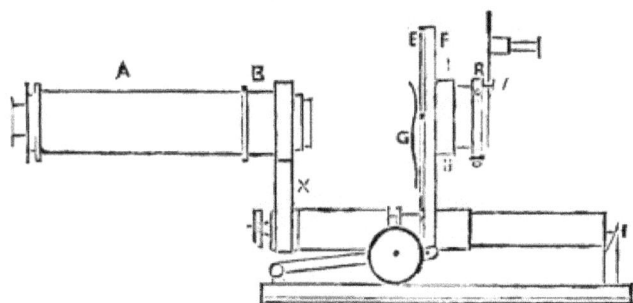

The accompanying figure represents a slightly modified form of Smith and Beck's popular monocular microscope. The method adopted for the inclination of the body is susceptible of improvement, but up to the present we have seen few stands to which the same observation is inapplicable.

The Tube A carrying the eye-piece, screws into B, which is permanently attached to the arm X carrying the objectives. The coarse and fine adjustments are conveniently placed, especially for photography, and their performance sufficiently perfect for most purposes.

The stage has been considerably modified, as the original was found to be too thick for investigations requiring very oblique light.

A thin circular brass plate E covered with blackened glass, revolves on F in the optic axis of the

instrument, on its upper surface two thin brass clips
G are fastened, these hold the slide in position and
allow of its movement in every direction in the plane
of the stage.

A brass tube H, lined with cloth, screws into the
centre of the stage, and holds the sub-stage apparatus in position, this can be removed when very
oblique light is required.

A diaphragm I, attached to a short wide tube with
a thick rim R, slides up and down in H, the motion
being easily imparted to it on account of the cloth
lining. A line divided into sixteenths of an inch, is
cut on the side of the tube at right angles to the rim
R, this enables us to tell the distance between the
top of the condenser and the object by calculating
the number of divisions between the rim and edge
of the other tube H.

This piece of apparatus, when made in the manner
about to be described, and which we strongly recommend, fulfils all the requirements of the most expensive sub-stage arrangements.

Fig. 16 is a diagramatic representation of it in
section.

FIG. 16.

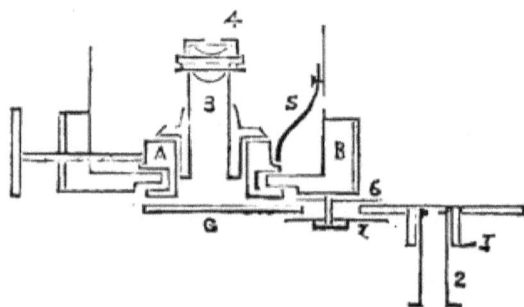

In the rim R two screws are fitted, (only one is represented in the figure for the sake of clearness) at right angles to each other, opposite to them in the interior of the tube a pretty strong spring S is attached.

By this means the ring A which fits on the projecting internal flanges of the rim R, can be moved in any direction in the plane of the stage by the screws, as the spring pressing against it on the opposite side compels it to follow their motions.

The diaphragm plate G is perforated with eight holes varying from one half to one sixteenth of an inch in diameter. In its centre is another aperture quarter of an inch in diameter. Two thin plates half an inch wide, through which the screw S fig. 16 passes, are placed above and below the diaphragm before it is fastened in position, thus allowing it to be displaced nearly quarter of an inch in its own plane, and parallel to that of the stage, quite independently of the remainder of the sub-stage apparatus.

Any one possessed of ordinary skill and a few tools for cutting sheet brass, will find little difficulty in applying this principle to existing instruments.

From what has been previously mentioned with regard to the resolution of certain structural peculiarities of microscopic objects, it follows, that their accurate delineation depends as much on the corrections of the condenser through which the light passes to the object, for achromatism, and on the size of its angular aperture, as it does on the perfection of the objective in these particulars.

It must be remembered however that a condenser is not necessary when using the one inch or lower power, but is indispensable with those above it.

A good achromatic condenser of the French form, consisting of three powers, screwing on top of each other will answer the purpose.

A much better form is Powell and Leland's new homogeneous immersions; whichever is used, it should be fitted in the centre of the ring A so as to be removable at pleasure, and with the face of the uppermost combination flush with the top of the substage tube, to allow of its close aproximation with the under surface of the glass slip on which the object is mounted.

Besides the condenser itself being freely removable from the ring in the substage, the three combinations which compose it should be easily unscrewed from one another, as on no account must the power of the condenser exceed that of the objective, we therefore require no condenser with the one inch, while with the quarter two and with the eight all three combinations are necessary.

If the condenser is not an immersion, a pin hole cap similar to that figured in section, 4, fig. 16, will be found an extremely useful addition placed on the top combination, when using the eighth and higher powers; it not only prevents the passage of any rays except those actually concerned in the direct illumination of the object and the subsequent formation of the image on the focusing screen, but also if brought into view with a low power objective enables us to make the optic axis of the condenser coincident with that of the objective, and therefore with any other substituted for it. This will be more fully explained when treating of the arrangement of the microscope prior to taking a photograph.

This pin hole cap is easily made as follows.

Procure a piece of tubing, that shall fit the top of the condenser accurately, lead will do, but brass is to be preferred. Cut it so that it projects slightly above the surfaces of combination, and while in this position press a piece of sheet zinc, previously cut to fit it, down into the tube till it just touches the glass.

Having filed down the part of the tube projecting above the zinc, we find the centre of the latter and pierce it with a hole about the diameter of an ordinary small pin. The cap is now blackened both externally and internally, with oxide of copper or any other material giving a dull black.

It has been found that the nearer the diaphragm approaches the back combination of the condenser, the more brilliant the illumination. The following arrangement which enables us to place it in immediate contact, besides possessing several advantages in practice, lessens the spherical aberration and improves the illumination, though it decreases the angle of aperture.

To the largest hole of the diaphragm a short tube, 1, fig. 16, lined with cloth is attached, another tube 2 of sufficient length to reach the under surface of the condenser, slides within it, at the upper extremity of the inner tube and a short distance from its ori-

fice a slight rim projects into the interior, on which
brass discs having different sized central apertures
may be fitted.

Another tube E carrying a disc of blue glass may
be made to slide within the tube 2. If the tint of
this glass is properly chosen it obviates the necessity of using a part of the spectrum to produce
monochromatic light.

The intensity of the colour is increased or diminished by increasing or diminishing the distance between the blue glass and the aperture of the slot or
back of the condenser.

The advantage of this sub-stage arrangement
is that no light can reach the object or enter the objective except through the diaphragm and condenser
themselves; the ease and simplicity with which the
several adjustments are made, and the possibility
of adapting it to any existing microscope stand.

ILLUMINATION.

The best light for Micro-photography is sunlight,
but the difficulties of its employment in our climate,
more than counterbalance the excellence of its results.

We shall give a description of the apparatus necessary for the benefit of those who can afford sufficient
time during daylight for Micro-photography, or who
may be more favourably situated atmospherically.

It should be borne in mind that the actinic power

FIG. 17.

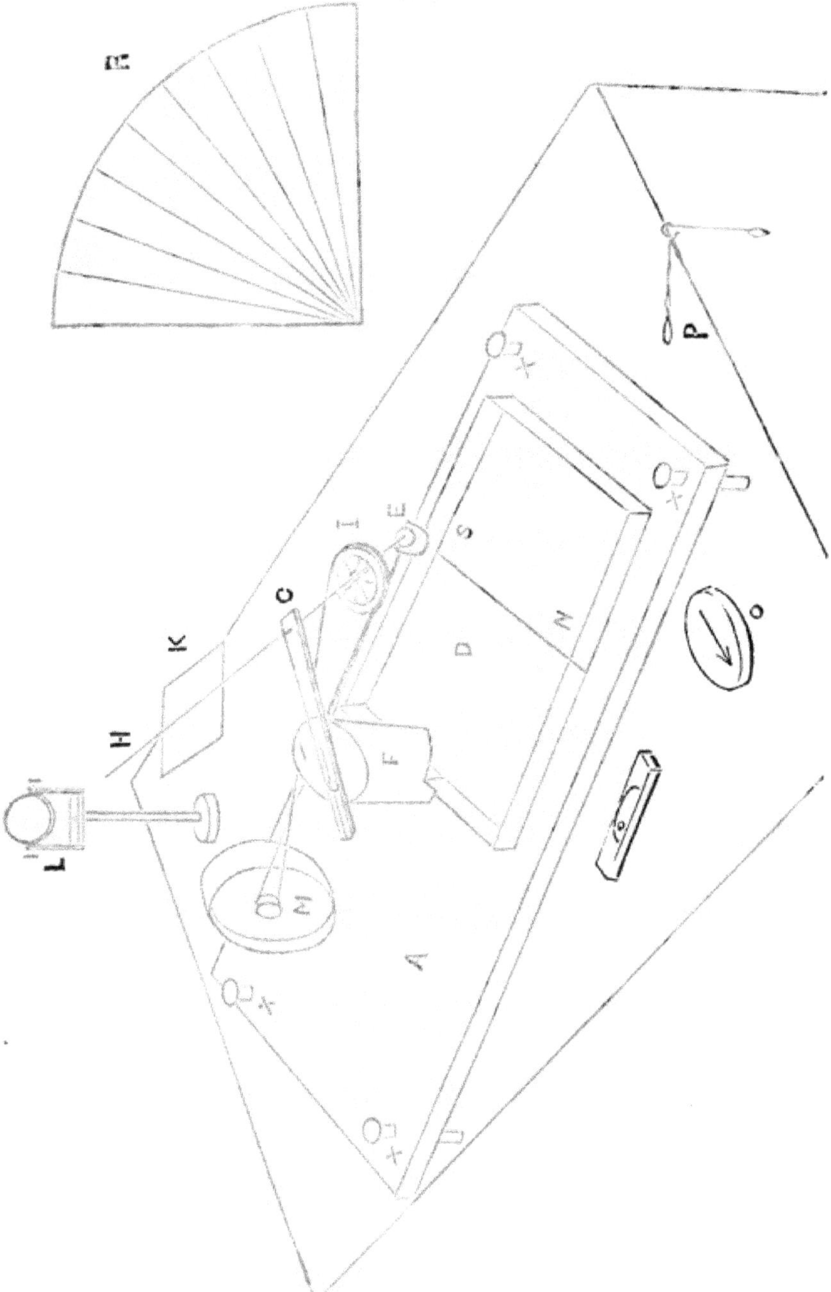

of the sun is subject to considerable variations, due
to its elevation above the horizon; this depends on
the hour of the day and time of the year, the greatest intensity being about noon in midsummer, while
at the same hour in midwinter it is about one third
less. No fixed rule can be given for exposure, its
duration therefore must be left to the judgment and
experience of the operator.

The rays of the sun may be thrown on the condenser of the microscope from a mirror moved by
hand. This is inconvenient and requires an assistant.
Some means of keeping the light in position is almost
a necessity, and for this purpose several instruments
have been invented, the cheapest being designed by
Stoney, and lately modified by Spencer of Dublin.
As all these instruments are expensive, we shall
describe a form of Heliostat very easily made and
sufficiently accurate for the purpose.

A thick well-seasoned oak or mahogany board
about six inches wide and twelve long, A, fig. 17, is
fitted with levelling screws, xxxx, at its four corners.
On it a small French clock M is placed, a boxwood
wheel one inch in circumference being previously
attached to the hour hand axle.

A sheet of white paper, with a line NS drawn
across it, is gummed to the back of a square of thick
plate glass, D, the whole being fastened to A, as in
the figure.

A ball and socket joint is fixed at E, the ball
having a hole half way through it, the orifice being

on a level with the top of the glass plate, and in a direct line with NS.

A wooden pillar F, the top being cut obliquely is placed as shown; on it is fastened a thick brass plate, by a binding screw, B, passing through the slot at one extremity, the other extremity being perforated to receive a knitting needle, H.

This Knitting Needle must be perfectly straight, and work very smoothly in E and G. On it between E and G a boxwood wheel, I, four inches in circumference is fixed, and above G a small mirror K.

A motion four times as slow as that of the hour hand of the clock is imparted to the needle and thus to the mirror, by a silk cord passing round the box-wood wheels M and I.

An elastic band and a small hook is the best means of keeping the cord tight, it should always be detached when the instrument is not in use.

Another and larger mirror, L, mounted to allow of its rotation in all directions, and elevation to any required height, is placed in front of the mirror K and generally on its left hand-side.

A stout card R whose sides are the same length as the needle, is divided into degrees, minutes, and seconds.

A small plumb line, P, a spirit level and a compass complete the apparatus.

Before describing the adjustment of the heliostat, it is necessary to call to mind the following facts.

The sun moves apparently in a circle of which the earth is the centre and the pole PP' the axis, fig. 18.

FIG. 18.

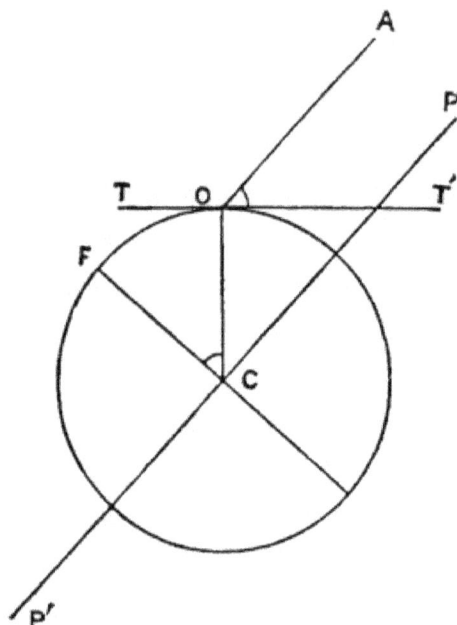

At any place O on the earth's surface a line AO
drawn pointing to the pole will be parallel to PP', as
the pole is infinitely distant.

If OC be now drawn and TT at right angles to it
and tangent to the place, the angle AOT', *i.e. the
altitude of the pole* = the angle OCF, *i.e.* the latitude
of the place.

Now since the sun completes a revolution in
24 hours it is evident that a mirror fixed at an
angle with the horizon equal to the latitude of the
place, will point to the pole, and if made to re-

volve by clockwork once in 24 hours will always face
the sun. If instead of facing the sun it is required to
throw its reflection on a given point, we must then
employ another mirror to receive the reflection from
the one looking towards the sun, and give the latter a
motion four times as slow to enable it to keep the
reflected ray in position on the second mirror. It is
on this account the boxwood wheel on the needle is
four times the circumference of that on the hour
axle of the clock.

When adjusting the heliostat it is best to fix it
permanently in position. The window of some room
looking towards the south and devoted entirely to
this purpose, should be chosen.

The stand is placed on a firm support at the window, (which should be open), so that the line NS
points north and south, this can easily be done by
means of a compass laid on the glass plate D over NS
The levelling of D being carried on at the same time,
by means of the screws XXX and a small spirit level.

The plumb-line is attached to the top of the needle
H, and the brass plate G moved till the lead of the
plumb-line hangs directly over NS. At the same
time the card R (which should be very accurately
made), is placed with its upper extremity at E, the
lower side being parallel with NS, and the needle H
made parallel with the line corresponding to the
latitude of the place on the card. H now points
to the pole.

During these adjustments, the level of the apparatus, and the north and south positions of the line

NS should not be disturbed. The position of the
needle should finally be tested by seeing that the
plumb line hangs directly over the N extremity of
NS. The silk cord is placed round the wheels on
the clock and needle, so that the mirror revolves in
the direction of the sun's motion when the clock is
set going. The sun's rays are directed on the mirror
L by the mirror K, and as L can be moved in all
directions, the reflected rays may be made to occupy
any position.

This instrument has been described for the benefit
of those who have the time and ability for the manufacture of their own instruments, and because it
cannot be purchased ready made. It was found
sufficiently accurate for exposures of moderate
length, but like all instruments of the kind it requires considerable care and practice to adjust.
When once fixed, its position should be left un-
changed, as a separate adjustment every time it is
required will be found very troublesome.

Stoney's heliostat can be purchased for five
pounds. It is a very reliable instrument and possesses the great advantage of only once reflecting
the incident ray.

Artificial light has so many advantages over sunlight that its employment is almost universal in
micro-photography, Swan's electric lamp is the best
form, but as an ordinary paraffin lamp gives excellent
results, and the arrangements are similar for both,
we will confine ourselves to a description of the latter.

Fig. 19 shows a simplified form of that used by Dr. Dallinger and described in the Journal of the Microscopical Society.

To the back of an ordinary tin paraffin lamp A, a board is attached to give rigidity; this board is fastened by screws to a block which can be raised or lowered on the pillar D by a rack and pinion movement. A lateral motion is communicated to the pillar D by means of the screw F.

FIG. 19.

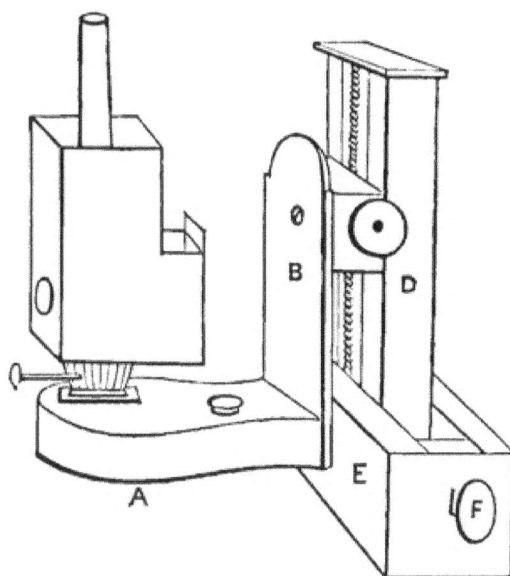

The lid of an oblong tin box about one inch wider than the chimney of the lamp, has a hole cut in it of sufficient size to allow of its fitting tightly on the small flange below the tongues of the burner, and as near as possible to the side of the lid intended for the front, that for the back being turnd down horizontally.

Another round hole two inches in diameter, is cut in the front of the box, at such a level that the flame may occupy its centre.

In the middle of the opposite side a slit three inches long is made, the upper extremity of which is cut again at right angles; the flaps thus formed are turned outward and a plane silvered mirror fastened between them.

The spaces left at the top and bottom of the flaps, allows a current of air to circulate between the mirror and the chimney of the lamp, and prevents their cracking.

An ordinary plano-convex condenser, its plane surface being turned towards the lamp, is placed about one inch from the latter, (when high powers are employed), and a stand to hold blue glass between this condenser and the microscope, The stand is unnecessary if the arrangement for obtaining a monochromatic light mentioned on a previous page has been carried out.

When using Swan's incandescent lamp, the same method of preventing extraneous light from reaching the microscope, and condensing the available portion on the object is employed. All batteries and apparatus for working it are best purchased, with regard to this full information will be supplied by any leading electrical firm.

The Camera and other Apparatus.

As the camera plays a very important part in micro-photography, we recommend the reader, unless he is a first rate mechanic, to purchase one; great care being required in its manufacture.

It should be quarter plate size, with bellows body, the interior lined with black velvet. The focussing screen which should consist of the finest patent plate (oiled), must be carefully adjusted to occupy exactly the same position as the sensitized plate.

It will be found convenient when working with dry plates, to have one or two double dark slides. Several plates can then be exposed consecutively, and developed together, by this a great waste of time and solutions is avoided.

The front, to which the lens is attached in ordinary photography, should slide in groves fixed to the body of the camera, we can then substitute for it a conical bellows, similar to that in Fig. 28, also lined with black velvet.

To the end of the conical bellows a tube an inch long is attached, this fits into the short tube of the microscope smoothly and accurately, its interior must be blackened, or better, covered with velvet, to prevent reflection from the sides.

The distance of the focussing screen from the front of the camera should be regulated, by a rack and pinion or screw motion, and when the desired position is obtained, some means of fastening it is necessary.

The plate carrier or dry back should be easily substituted for the focussing screen, as the slightest shake might destroy an arrangement of the microscope and its accessories, that consumed a considerable time in preparation.

The camera, when these additions are completed, is fastened to an oblong board, the same width as itself, but projecting about two inches beyond it, both back and front. To the corners of this board Levelling screws are attached which enable the focussing screen to be made coincident with the optic axis of the microscope, and the tube on the bellows to be placed in the most favourable position for adjustment to that instrument.

The greatest difficulty that besets us, especially when using artificial light, is the impossibility of focussing minute markings on the screen. This is due to three causes, first the coarseness of the surface on which the image is formed, second the minuteness of the image itself and thirdly *want of light*. The first is irremediable as nothing has yet been found, of sufficient surface opacity and fineness of texture, to take the place of patent plate.

The second may be remedied by magnifying the image. A focussing glass answers this purpose, an excellent one being made as follows :—

With a piece of stout cardboard make a tube four or five inches long to fit one of the eye-pieces of the

microscope, and blacken it inside. In the opposite
end fix a plano convex lens about three inches focus,
its convex side being turned towards the eye-piece,
or preferably a universal screw to receive an objective of about the same power. Make another tube
six or seven inches long to fit over the first and the
focussing glass is completed.

The end of the outer tube being placed against
the screen, its roughened side is brought into focus
by sliding the inner tube within the outer. When the
best focus has been obtained, which may be done by
fastening a very small fly or other minute object on
the focussing screen, the inner tube is marked with
a circle where it joins the outer, thus enabling it to
be placed in the same position when again required.

The third difficulty, namely, want of light, cannot
be overcome, as any increase in quantity of the light,
decreases the visibility of the structure.

The stand intended to support the microscope and
its accessories requires to be made as firmly as possible. The following is a description of the form
finally adapted by the writer after more than ten
years' experience, during which time scarcely a month
elapsed without the trial of some fresh device. It is
still far from perfect, but we hope, its simplicity
will in some degree atone for its numerous defects.

A beam ten inches wide two inches thick and six
feet long, is sawn into three separate portions, three,
one, and two feet long respectively. On the upper
surface at each side flush with the edges, two laths
one inch wide, half an inch thick, and the same

length as the beam, are fastened, and divided into
three parts, the same way as the beam. These not
only prevent anything placed on the stand from
rolling or being knocked off, but if divided into
inches and half inches, enable the apparatus to be
placed in any position they may have formerly
occupied on the stand.

Parts 1 and 2 are joined together on their under
surface by two strong hinges, let in flush with the
wood. 2 and 3 are joined on their upper surface, in
the same way where the laths meet as in fig. 23.

Along the sides of the beams ten bolts are inserted
projecting one inch and a quarter from the surface,
and six inches distant from each other. Their proper
position is best determined by first fixing two, three
inches on each side of the centre of plank no. 2, or
three inches from where it joins 1 and 3; then in-
serting the remaining eight ten inches from these
and from each other at four equal distances on planks
1 and 3. This description will be easily understood
by a reference to the figure.

A board three feet long, two and a half inches
wide, and one inch thick, B fig. 23, has six holes
drilled in it, six inches apart, the last holes on either
side being three inches from its extremities. This
enables it to be screwed on to any six bolts at the
same time, thus keeping the several parts of the
beam steady, in the various positions we are compelled to place it, owing to the different methods
adopted to meet the requirements of every case.

The first division of the beam I, fig. 23, is intended to support the oblong board to which the camera is fastened; a binding screw and a brass slip working in a slot cut down the centre of the beam attaches them to each other.

On the second division, 2, a platform of sufficient height to enable the microscope to be adjusted to the camera is erected, the microscope being previously firmly attached to a separate board, sliding to and from the lamp, between two grooved slips on the top of the platform. This should also be clamped in position by a binding screw.

Owing to the different forms of microscope stands it is impossible to adhere to any fixed rule with regard to the method of attaching the microscope to the platform. It must, however, be as accurately and firmly fastened as possible, while its motion to and from the lamp must take place without disturbing the centering of the different instruments.

A thick board the same width as the stand and about twelve or fourteen inches long, is fastened at right angles to plank No. 3 on the extremity behind the lamp; an aperture six inches square being previously cut in it, on a level with the top of the platform to which the camera is attached. This aperture enables direct sunlight to be used, with the apparatus in the horizontal position.

A division about quarter of an inch wide, runs along the centre of the plank to within two or three inches of its extremities, this enables the lamp, bull's eye condenser, and sulphate of copper cell to be firmly clamped in position by binding screws.

The stand can be placed at any angle, by shortening or lengthening the camera legs attached to the bolts on the extremity of plank No. 1. They may be detached when not required.

A reference to figures (23, 24, 25), will enable us to understand why the arrangement is so complicated. It will be seen that five different methods of placing the apparatus are adopted, and the stand made to suit them all. A full description is reserved for another portion of the work, devoted to the particular arrangement of the stand, with certain objects and objectives.

Those positions in which the apparatus is horizontal should always be chosen if possible, not only because it is the most convenient for manipulation, but on account of the superiority of the illumination.

The perpendicular position is only necessary when the object is fluid, and the positions between these when oblique light is required.

An iron rod three feet long, sliding in a tube beneath the oblong board supporting the camera, is fixed so that a wheel, attached to the extremity next the microscope, may always occupy the same position beneath the fine adjustment, independently of the distance of the camera from the platform. A silk cord fastened round this wheel and the fine adjustment, enables an observer, by turning the milled head at the opposite end of the rod, to focus the image on the screen of the camera when the adjustment is beyond reach. An assistant, is then unnecessary as no matter how intelligent he may be, it will be found more satisfactory to have the focussing under our own control.

This method is a modification of that adopted by Mr. Woodward when photographing without a camera, under which circumstances we never use it, having found the following arrangement much more satisfactory.

A cork with a slight rod attached to it by two wire holdfasts, is slipped into a tube fastened to the fine adjustment of the microscope; this enables us to act directly on the fine adjustment without the intervention of any cord or wheels; unfortunately its use is limited to those methods in which the microscope is not attached to the camera.

When sunlight is the source of illumination, there is great danger of the object and objective being destroyed by the heat rays focussed on them by the condenser. This may be partially avoided by putting the condenser a little out of focus, or we may entirely eliminate them by the interposition of a plate of alum between the heliostat and condenser.

The best method, however, is to remove the blue glass, and place a cell having parallel sides, and filled with a solution of ammonia sulphate of copper in a similar position to that recommended for the alum plate. The cell is made as follows :—

Fig. 20.

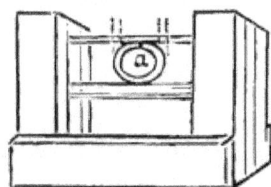

A piece of round india-rubber placed between two
quarter plates, is made to occupy the same position as the dotted lines in figure (20). A board
slightly narrower than the plates and india-rubber,
has a piece cut out of it $4\frac{1}{4}$ inches wide and about
the same length; four thin strips are fastened along
its edges and the plates with the india-rubber forced
down between them; when this has been done the
cell is filled with the solution, and the india-rubber
pressed down so as to complete the ring and prevent
evaporation.

This cell may with advantage be used on all occa-
sions instead of the blue glass.

A bull's eye condenser of about four inches focus and
one and a half inches diameter, is indispensable.
It should be mounted as shown in the engraving
the base being made of lead to increase its stabi-
lity.

Having finished our description of the apparatus
required in Micro-photography, we wish finally to
impress on the reader the necessity of making them
as well and accurately as possible, and not to attempt
their manufacture himself unless he is a good mechanic.

We now pass to the arrangement of the dark
room. The outlay is very small and will amply re-
pay the photographer.

Any spare room may be selected for its erection,
preferably on the ground floor, near or if possible in
the apartment set apart for micro-photography and
the instruments used for that purpose.

About six feet from one wall and twelve from the other, a post A, fig. 21, reaching from floor to ceiling is erected, two more posts B and C are placed opposite to it against the walls, and two or three long laths are fastened at equal intervals horizontally between A and B.

Fig. 21.

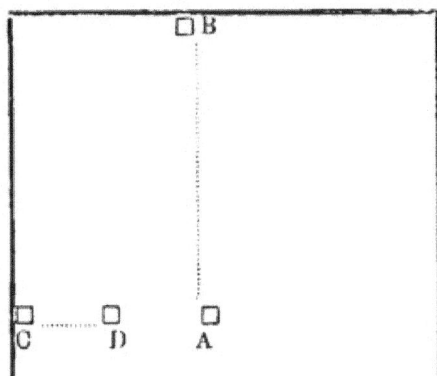

Half way between A and C another post D is fixed, and about seven feet from the floor a cross beam between A and D; a wooden frame fitting accurately in this interval is hinged to A, and forms the basis of the future door.

The whole is covered with two thicknesses of old newspaper on the outside, care being taken that they are firmly pasted to both floor and ceiling; when dry the part covering the frame-work of the door, is cut along the edges and the door opened.

The inside is now covered in the same way, and when dry the process repeated inside and outside with the thickest tarred paper. A black curtain is hung over the door. The apartment will now be found to be completely light tight.

The advantage of the paper walls is that a window can be cut in any convenient part, and a couple of sheets of ruby glass a little larger than the orifice fastened over it, by a few strips of orange paper pasted round their edges.

The best position for the window is four feet from the floor in the longest side, and the same distance from the wall opposite the door.

Underneath the window a wooden trough lined with lead, about three feet long, eighteen inches wide, and three inches deep, is fixed; it should have a gradual incline towards a hole in its centre, where

Fig. 22.

the waste pipe is attached, which should pass out-side both apartments if possible.

The top should be levelled and a sliding frame-work, or a board with holes placed upon it, as in (fig. 22). If no better water supply can be obtained a barrel capable of holding a sufficient quantity and having a brass rose cock near the bottom, may be fixed on a stand, about three or four inches higher than the trough and on its left hand side.

Some shelves to hold stock solution, etc., are placed against the wall opposite to the window, a small shelf being fastened above and below the window itself to hold the bottles and dishes in actual use.

On the right hand side of the trough a square box is fixed just deep enough, to allow the flame of a lamp, with a chimney of ruby glass, to appear above it.

The best method of preventing any actinic light from the lamp reaching the sensitive plates, is to cut two holes in a large American meat tin, one on a level with the flame, the other in the top. In the latter a crooked chimney, the same as is used for a magic lantern, is inserted; and in the former a piece of ruby glass. The whole is placed over the lamp in the box, (the sides of the tin near the bottom having a few holes previously made in them, to allow of the free entrance of air).

One or two cupboards are useful, they should be large enough to hold tin boxes for plates, sensitive tissue, etc.

It will be seen by the figure that, in the unoccu-
pied portion of the paper wall next the door, about
three feet from the ground, a window twelve inches
wide and about four or five feet long, has been cut,
a board two feet long the same width as the window
projects from its lower end into the room, where it
rests on a support which keeps it perfectly level.
Two laths or posts are erected at the dark room extremity of this platform, and the whole covered over
with newspapers and tarred paper, the same way as
the dark room itself; a reference to fig. 26 will explain this better than any words can do.

A recess is thus formed in the wall outside of the
dark room. Inside the dark room in the side walls
of this recess, two windows about a foot square are
cut, and covered with sleeves of black cloth, fastened
at their open ends with indiarubber.

In the partition between the dark room and interior of the recess, between these two walls an orifice is made, about thirty inches in diameter also
covered with a sleeve of black cloth.

The legs of the stand described in the previous
chapter are taken off, plank no. 1 from which the
camera is detached folded beneath 2 and 3, and
placed in this recess with the microscope towards
the dark room, all the necessary manipulations to be
described in future pages, are carried on by the arms
passed through the two sleeves situated at the side
of the recess. The third sleeve is fastened round the
tube of the microscope, and the fine adjustment
screw, the milled head being removed, passed

through a small orifice in it; the rod and cork before described are now put on, and all arrangements for working, without leaving the dark room, completed.

When possible we strongly recommend this latter arrangement.

CHAPTER IV.

Mounting and Preparation of Objects.

The present chapter might be considered superfluous but for the fact that objects designed for Micro-photography require different treatment from those intended for ordinary microscopic examination. In addition to this it is right that those who have obtained even a small amount of knowledge in practical science should impart it to others, and thus prevent the disappointment and loss of time, consequent on a dearly bought experience.

The usual method of mounting microscopic specimens, is after preparing them by some of the ways mentioned farther on, to place them on plate-glass slips three inches long and one wide, then after the addition of a small quantity of a suitable preservative medium, they are covered with a square or circle of very thin glass, which is fastened in position either by the medium itself, or by a ring of cement painted round its edge.

The disadvantages of this method will be apparent when the advantages, of the one we advocate, are stated. We never expect its universal adoption, because the old plan gives greater scope for artistic display, and the manufacture of a kind of article whose external appearance is more likely to catch the eye of the public.

Our method is as follows. After the usual preparation, the specimen is placed on a square of thin glass, (which may be fastened to an ordinary wooden slip or not, we prefer not), and covered after the application of a suitable medium with another square of the same size, both if not fastened by the medium itself are cemented by passing a brush charged with any of the usual varnishes round their edges. When dry they are either permanently inserted in a boxwood slip A, fig. 23, or what is much more convenient, placed when required for examination within or on one, by any of the methods shown in the figure being adopted.

The advantages claimed for this plan are, convenience of storage, postal transmission, etc.; possibility of examining both sides of the specimen; a high power objective sustains no injury if it happens to be driven through the preparation; the aberrations caused by the thickness of the glass slides usually employed, are removed; when using high powers the condenser can be brought close to the object. The examination of objects with oblique light is facilitated, and their definition wonderfully improved; and even with direct light, markings can be seen which defy resolution, if the object be mounted in the usual way; and finally one slip alone is necessary for the reception of any preparation requiring examination.

Many objects, among which may be mentioned, scales of moths and butterflies, etc., require no previous preparation, all that is necessary after placing them between thin glass, is to pass a

FIG. 23.

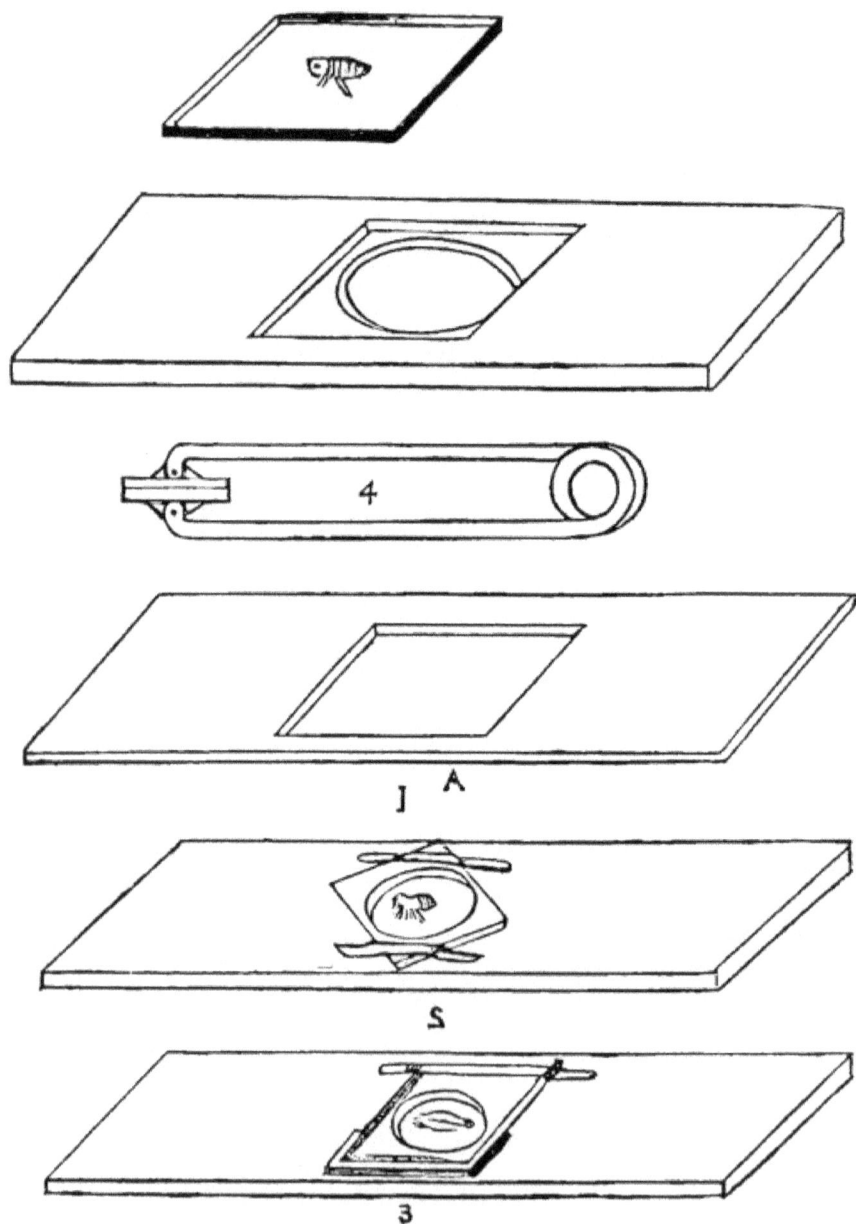

brush charged with asphaltum round their edges. This operation is facilitated if the thin glass is held in the clip 4, fig. 23. The cement should be as thick as possible to prevent its running in between the glasses. Preparations liable to injury from pressure, will if thin, require a ring of cement painted round one or both glasses and allowed to dry before the preparation is placed between them, and if thick a ring of paper or zinc answers the same purpose.

Whole insects or opaque parts of the same are rendered transparent by soaking in turpentine for two or three days, and afterwards mounted in balsam.

The balsam must be first evaporated by gentle heat, and when it has attained the consistency of resin, it is dissolved in an equal quantity of bensol and kept in a bottle with an air-tight glass cap for use.

Balsam should only be used when nothing else will answer, as its high refractive index renders the finer details of certain structures invisible, besides the yellow colour of this medium causes as much diminution of the actinic power of the light, in the illuminated parts of the field, as the shaded portions of the object itself; thus markings, plainly visible in the microscope, fail to make any impression on the sensitised surface of the plate.

Damar, being whiter in colour and its refractive index less, gives better results. As a cement it is much superior, not becoming brittle like balsam after some time.

Minute algœ infusoria, etc., require no previous preparation, but may be at once placed in glycerine and camphor water, by this means their original colour is preserved and the growth of mycelium prevented. It should be remembered that glycerine cannot be employed as a preservative for objects containing carbonate of lime, as it is a solvent of that substance. Dr. Carpenter recommends its saturation by pieces of marble as a preventitive. It is much safer to mount such specimens in distilled water to which a minute quantity of creosote has been added.

For further information as to the relative merits of different media, we refer the reader to the works of Klein, Carpenter, Beale, etc., mentioning at the same time, that those recommended in these pages will be found generally useful.

Few more interesting objects present themselves for examination than the polycystinœ, their variety of form and beauty of outline, rendering them equally a source of delight to the observer and of admiration to the artist.

Their distribution is very general, beautiful forms being found in chalk, limestone, and many other deposits. Large quantities are contained in the sand removed from sponges, this may be obtained from any chemist or sponge merchant without payment.

The best method of separating them from the dirt and debris, with which they are mixed, is to wash the deposit in a large quantity of water with frequent shakings for an hour. After standing the same

length of time, the fluid is removed without disturbing the sediment at the bottom of the vessel. This
is repeated till the washings become colourless.

A small quantity of the sediment is transferred by
a pipette to a piece of thin glass, and the fluid
allowed to evaporate; the evaporation may be ac-
celerated by heat.

The thin glass is next fixed in the slip and placed
on the stage of the microscope. A low power objective adjusted and the coarser uninteresting particles removed by a bristle from an old shaving
brush, fastened in a wooden handle. These bristles
are nearly always split, and when pressed on the
glass their ends separate; the object is allowed to
occupy a position between them ; on removal of the
pressure the bristle closes and the object if valuable
may be transferred to another slide, or if not, dis-
carded.

It may be perceived that two methods are adopted,
either the removal of the debris from the original
slide, or the transference of the objects to a fresh
one, the former being employed when the objects
are in excess of the debris, the latter when the
opposite is the case.

Whichever plan is adopted, a perfectly clean cover
glass being seized by a small pointed forceps, a
drop of damar is placed near the edge of the side
farthest from the forceps, this side is brought into
contact with the corresponding side of the slip on
which the objects lie, the other is allowed to sink
gradually to its position, in such a manner, that the

fluid goes in a wave from one side to the other, carrying all air bubbles before it.

Should air bubbles make their appearance, one edge of the cover glass may be raised with a fine needle and an effort made to bring them to the side, as they are inclined to adhere to the edge of the fluid in its passage outwards. They may also be removed by placing the preparation under a small air pump. This latter piece of apparatus will be found very useful for removing air bubbles from structures liable to injury by displacement of the cover glass.

When faraminefera are contained in limestone or other hard substances, sections of the substance sufficiently thin to be transparent are required, instructions for cutting them will be given farther on.

We now pass to the preparation of the diatomaceœ, they are closely allied to the polycystinœ and are found in the same deposits. Beautiful forms have been obtained from guano, infusorial earths, sea dredgings, and stomachs of molluscæ. They are almost ubiquitous, every road and ditch being filled with them, especially those containing brackish water. They may be obtained from the latter tolerably free from admixture, if proper care be taken, during their collection, to leave the bottom mud undisturbed.

If desirable, they may be mounted in their original condition in distilled water and camphor, but if the silicious skeletons alone are required, the gelatinous envelope must be removed by boiling in nitric acid, the sediment well washed till all trace of acid has disappeared, and the residue mounted dry.

When contained in other deposits two ounces of the mixture are placed in two gallons of water, after violent shaking allowed to stand two hours, and the supernatant fluid removed. This process is repeated till the water is no longer coloured. As much fluid as possible being removed, the deposit is treated with strong hydrochloric acid, and allowed to stand for an hour. The acid is then poured off and a fresh supply added, the whole being boiled for one hour and washed in the manner first described, till all traces of acid are removed. The process is repeated with strong nitric acid, and after subsequent washing to free it from acid, the deposit is placed in a tumbler full of water. Now comes the most trying part of the preparation.

The tumbler is allowed to stand five minutes, at the end of that time its liquid contents are transferred to another tumbler which we shall call number 2, number 1 being filled and shaken or stirred with a wooden rod, after standing five minutes its fluid contents are tranferred to a number 3, number 1 is again filled and stirred, left to stand for five minutes, at the end of which time the liquid in number 2 is transferred to a number 4, and the contents of number 1 put in its place; number 1 being again filled and left standing five minutes, the contents of number 3 are transferred to a number 5 and the fluid of number 1 put in its place; numbers 4 and 5 are allowed to stand a quarter of an hour, the changes being regulated in the other tumblers to suit this, by adding 6 and 7 to them and so on till we manage to collect 12 single tumblers, whose contents represent washings lasting from 5 to 60 minutes respectively, this process will take a long time as the washings multiply indefinitely; towards the end some of the tumblers will be found to contain all the diatomaceæ tolerably free from impurities. The labour is not so great as might appear at first sight, for every tumbler that has undergone twelve washings is put aside.

It has been recommended to mount diatomaceæ in bisulphide of carbon, as its high refractive index makes their markings more distinct. We have as yet had no experience of its efficacy in rendering their details developable on a sensitised surface.

Since the first discovery of the markings on the diatomaceæ their true nature has been the source of endless strife, and is at the present moment in as unsettled a condition as ever. I remember having experienced immense delight many years ago, when for the first time I perceived that the dots on some of the coscinodesceæ consisted of a coarse and fine variety, and that I could reproduce exactly the same appearance by placing two pieces of glass covered with hemispherical elevations, within two or three inches of each other. Since then my ardour with regard to the investigation of the structure of the diatomaceæ has cooled, and I have made no further attempt to prove how far the foregoing experiment is an explanation of their appearance. My

belief is that they are hemispherical tuberosities, in
the diatomaceæ proper. Their markings possess little
interest now, since they are allowed to have no value
even as tests, for resolving power of objectives; in
fact the investigation of their structure has led to
such a mis-application of talent and waste of time,
that the majority of workers in science have handed
this department over to the amateur.

Botanical preparations require considerable care in
their preparation, and as a rule are best examined in
the fresh state. A few, such as sections of stems,
etc., may be mounted in glycerine and a thick solution of gum in camphor water. Opaque objects as
sori of ferns, may after soaking in turpentine be
mounted in balsam; but others require a process of
maceration, staining and dissection, which will be
best understood by a reference to some text-book on
botany.

Pathological and histological objects are prepared
and mounted by the following methods.

When the intimate structure of tissue requires investigation, it should be obtained as fresh as possible
and a small portion immediately teased out with
needles, in a little serum or glycerine on a glass
slide. It will be found advantageous to insert the
needles into light wooden handles with a pliers, and
to place a piece of black cloth or white paper underneath the slide during dissection. When the dissection is finished, the tissue is steeped for three minutes
in a ½ p.c. solution of chloride of gold, or a small
quantity of the toning solution used for prints
diluted with three or four times its weight of dis-

tilled water. After an immersion of three minutes it
is removed from the gold solution and exposed to
the light in a little glycerine, until it assumes a
violet gray shade. This will take about twelve
hours. A ½ p. c. solution of nitrate of silver may
be used in the same way, and is extremely useful
for histological work especially the demonstration of
the endothelial linings of serous membranes.

These preparations are finally mounted in glycerine, and kept in the dark when not under examination.

The majority of pathological products require to
be cut into thin sections, before their examination is
possible. For this purpose they are placed while
fresh in Muller's fluid, which consists of—

> Bichromate of Potash, 2 parts.
> Sodium Sulphate, 1 part.
> Water, 100 parts.

This should be omitted if the sections require
staining with chloride of gold, or nitrate of silver.

The organ or tissue is cut into sufficiently small
pieces, and placed for three hours in a very thick
solution of gum. A little alcohol added to the gum
prevents putrefaction and allows it to be used for a
considerable time.

The instruments required for section cutting are a
razor and a microtome. The razor is set in a strong
wooden handle, it should have a good thick back,
both sides being hollow ground.

Rutherford's microtome consists of a brass table,

in which a cylinder containing a piston is inserted. The piston can be raised or lowered by a fine threaded screw. A box covered with gutta percha and projecting beyond the table, surrounds the cylinder; in its bottom is a drainage tube to carry off the melted ice and salt.

The microtome is attached to a firm table, the piston well oiled and the brass table polished; near at hand, we place the razor and a strop, as it requires sharpening after two or three sections have been cut.

A block of ice splintered by a sharp awl, is pounded as fine as possible and placed in an earthen-ware crock or other suitable receptacle. An equal quantity of coarse salt being put in a similar vessel.

The razor and a clean camel's hair brush are placed in a dish containing water to which a small quantity of alcohol has been added.

The cylinder is nearly filled with thick gum. One of the pieces of tissue or organ being submerged in it, the orifice is covered by an inverted wine glass, and the box filled with alternate spoonfuls of ice and salt. The mixture must be tightly rammed round the cylinder, care being taken not to stop up the drainage tube.

The freezing should proceed slowly, and only be carried far enough to cause the gum and tissue to assume a grizzly consistency. If they become white and icy looking it has been carried too far, and time must be allowed for their return to a proper condition before using the razor.

The rough top being sliced off and the razor well

F

wetted with spirit and water, the handle is grasped by the right hand the edge being turned from us.

A slight turn is now given to the screw to elevate the cylinder of gum above the level of the table, the ends of the fingers of the left hand are placed on the opposite extremity of the razor from the handle, the edge of the razor near the left hand fingers is applied to the near side of the gum cylinder, and passed through it from point to heel. The edge during this manœuvre being lightly applied to the brass table, while the back is slightly elevated above it.

The thickness of the section depends on the distance through which the screw is turned, if the latter be graduated, which is very easily done, the exact thickness may be regulated. To graduate the screw a measure is placed in the cylinder, and the number of turns required to elevate the piston a quarter of an inch counted. If it takes twenty-five, every revolution of the screw elevates the cylinder one-hundreth of an inch. A mark is now made on the milled head of the screw, and the end of the cylinder round which it works is divided (round its circumference) into five equal parts. The mark on the screw when it moves through one of these divisions raises the section one five-hundredth of an inch, a thinness rarely attained in section cutting.

Several sections being cut by this method without stopping to remove them singly from the razor, the razor with the sections on it, are placed in the dish containing the spirit and water; after a minute

or two and a little gentle agitation they will float off,
any that adhere being cautiously displaced with the
camel's hair brush.

When a sufficient number of sections are cut, they
are taken from the spirit and water, and placed in the
staining solution ; a flat tin spoon having a hole in
the centre of its rounded extremity and its handle
bent at right angles, will be found useful for this
purpose.

The best staining fluid for all preparations intended for microscopical photography is Beale's
carmine solution. Klein recommends the following
modification,

Powdered Carmine, 2 grammes,

rub up with a few drops of distilled water and add

Liquor Ammon. Fort., 4 c.c.
Distilled Water, 48 c.c.

Shake up well and place in a bottle secured with
an india rubber or greased cork to prevent evaporation. Glass stoppers are useless for this purpose and
should only be used for acids.

When required one drop is added to 10 or 20 of
water. If the tissue is very hard or has been kept a
long time in Muller's fluid, a stronger solution will
be required, and the vessel containing the staining
fluid with the sections must be accurately closed,
as its penetration depends to a considerable degree
on the quantity of ammonia. If however the sections are fresh one drop is sufficient and the greater
part of the ammonia should be allowed to evaporate

before the addition of water. The proper amount
of staining requires considerable judgment to determine, but the more dilute the solution, the easier it
becomes and the greater the differentiation of tissue.
When forming an opinion, the condition of the
tissue, as regards its age, (embryonic tissues staining very readily) its hardness, the alkalinity of the
solution, and the magnifying power required to
show it properly should all be taken into consideration. For further particulars we refer to Beale's
"How to Work with the Microscope," etc.; with
this caution that all blue and violet staining solutions
are useless, unless the operator has unlimited time
at his disposal for the modification of the photographs by various methods of intensification, after
their direct production.

If the sections are thin and intended for minute
investigation, they are removed from the carmine
solution by the flat spoon and placed for a short
time in equal parts of distilled water and alcohol, to
remove excess of colouring matter; then transferred
to glycerine and camphor water, where they may be
allowed to remain for a day or two, (covered with a
bell jar), before mounting, in gum and glycerine.
If thick and only required for very low powers, after
removal from the carmine, they are placed in *absolute*
alcohol, for twenty-four hours. When this period has
elapsed each section is taken up separately by the
spoon, both are firmly pressed between two folds
of blotting paper held between the finger and thumb,
(no danger of injuring the specimens need be appre-

hended, as they have been considerably hardened by the alcohol), and transferred to a saucer containing clove oil, in which they are left for two or three hours, covered by a bell jar as before.

The clove oil is removed from the section in the same manner as the alcohol, and a drop or two of damar placed upon it before its removal from the spoon. When it becomes transparent, which generally takes place after the lapse of a few seconds, it is floated on to the slide by the aid of a needle, and finally mounted in the manner already described.

The discovery of bacteria in diseased tissue, and the probability that a special form accompanies each disease, has excited considerable interest in the labours of Ehrich, Koch, and others, who have investigated the subject, and to whom we are indebted for the methods of rendering the organisms visible.

The following is a modification of the method adopted for demonstrating the presence of the tubercle bacillus in sputa or diseased tissue, and is an abstract from a paper read in the pathological section of the British Medical Association at Worcester, in August, 1882, by Dr. Heneage Gibbes. The following stock solutions are prepared.

    I. Magenta Crystals, grammes 2.
       Pure Aniline,   C.C. 3.
       Alcohol sp. gr. ·830, C.C. 20.
       Distilled Water,   C.C. 20.

Dissolve the aniline in the spirit ; rub up the magenta in a glass mortar ; add the spirit gradually while stirring, till all the colour is dissolved, then add the water slowly, still stirring, and put in a stoppered bottle.

> II. Saturated solution of chrysoiden in distilled water, to which is added a crystal of thymol dissolved in a little absolute alcohol.

This solution must be kept in a dark place.

> III. Commercial Nitric Acid, 1 part.
> Distilled Water, 2 parts.

The first two solutions must not be filtered.

The solutions are used as follows :—

The sputum is spread on a cover glass and allowed to dry, it is then passed three or four times through the flame of a spirit lamp, and left to cool. A few drops of solution, number 1, are filtered into a watch glass, and the cover glass, sputum downwards, placed on it, care being taken that no air bubbles are under the cover glass ; it is allowed to remain twenty minutes. It is then placed in the acid solution, number 3, until the colour is removed or about ten minutes. It is then washed in distilled water until the nitric acid is removed, and then placed in a little chrysoidin solution which has been filtered into a watch glass, for a few minutes. It must again be washed in distilled water, and the superfluous water drained off on filtering paper ; it is then placed in absolute alcohol and dried thoroughly in air. When dry it is mounted in Balsam in the usual manner.

Sections of hardened tissue are treated in the same manner, and the bacillus is shown by this method equally well in sections hardened in spirit and chromic acid.

We consider this an excellent modification of the older methods, and although the photograph of the bacillus anthracis in the frontispiece, was prepared by Ehrichs original method, the results obtained with the tubercle bacillus, by the modified plan, were superior.

Bone and all substances containing an organic basis impregnated with earthy salts may be softened by dissolving out the latter by prolonged maceration in dilute acid, their consistency after this treatment will allow of their being cut in the microtome without freezing, if embedded in parafin and wax.

Rocks, fossils, and bone when not softened, require entirely different treatment and considerable skill before they arrive at a condition suitable for microscopic investigation. A very slight sketch of the plan adopted is here given to enable those who do not possess a lapidary's workshop to investigate these substances under low powers.

A rough slice is cut from the substance with a small saw or by chipping ; one side is fastened with balsam about an inch from the extremity of a board ; twelve inches long, two inches wide, and half an inch thick. When the balsam has hardened, the upper surface of the specimen is flattened with a file, next rubbed down on a perfectly flat Arkansus stone; and finally polished with putty powder spread on a leather pad; a simple microscope is useful for watching the progress of the latter operation. When finished this side is fastened on the wood; the specimen filed down as thin as possible, and the foregoing process repeated.

If the specimen is very brittle it is best attached to a glass slip, and saturated with balsam, before commencing operations on the second surface.

The finished section may be now removed from the wooden support by soaking in turpentine, after which it is transferred to clove oil, and then to absolute alcohol. On removal from this the alcohol is allowed to evaporate spontaneously, and the section mounted dry or in balsam.

For the preparation of injected specimens and other processes, we refer the reader to the authorities already mentioned, reminding them of the fact, that violet and blue solutions are unsatisfactory for staining or injecting sections intended for photography, and also that only two methods of mounting, namely, in gum and glycerine, and dry, are required for any preparation within the domain of microscopy.

The secret of successful mounting is cleanliness; all solutions must be filtered before and after using; specimens, while in them, should be covered with a bell jar; and all media should be kept in wide mouthed bottles, closed by glass caps, not corks or stoppers.

The most convenient articles for keeping sections in during the staining and cleaning processes are large ointment pots; in lieu of these saucers may be used.

Glass rods should be kept in the bottles for removing small quantities of the media, to the slides.

The thin glass when purchased should be cleaned in dilute nitric acid, and kept for use in small jars containing alcohol. When required one is taken out, dried in an old linen or calico rag, and finally polished with a piece of soft wash-leather; this may be easily accomplished by folding the leather over it, and placing it when thus covered between the finger and thumb; a good deal of pressure and polishing can be done in this way without much risk of fracture.

All specimens when mounted must be kept in flat trays in a horizontal position, the trays themselves being placed in a box opening at the side and top to allow of their easy removal.

CHAPTER V.

PHOTOGRAPHIC PROCESSES.

FROM among the many processes used in photography we select the two best known, and most generally practised, referring the reader for further details to works devoted to that subject.

The wet collodion process will first claim our attention, not only on account of the priority of its invention, but because the gradations of the resulting image and the fineness of the deposit give a beauty to the results, hitherto unapproached by any other method. Unfortunately on account of the lengthened exposure required, its employment in Micro-photography with high powers, is only possible when sunlight is the source of illumination; but with low powers, and when an exposure of not more than ten minutes is necessary, it is generally applicable.

The size of plate required for Micro-photography is the carte-de-visite or quarter plate; the best obtainable are made of imprimatur glass.

If used before, for the production of a negative, the old picture can be removed from the glass by warm water and a soft rag, then washed in a strong solution of caustic potash to remove grease; dipped in water again several times, next in nitric acid, and finally left soaking in a dish of water acidulated with nitric acid.

When required, a plate is taken out rinsed in pure water and polished with a soft rag or piece of wash-leather, kept specially for that purpose.

The easiest method of testing the cleanliness of the surface is by breathing on it; if the moisture of the breath is evenly distributed, the plate is clean.

Collodion is difficult to prepare, but bromo-iodised collodion, of first-rate quality, can be purchased from any leading firm; and will give better results than that prepared at home.

A vertical glass bath, with a cover to keep out the dust, etc., will be found the most useful; a pure silver wire dipper, for immersing the plate in the solution, being purchased along with it.

The sensitizing solution placed in the bath is made as follows :—

> Recrystalized Nitrate of Silver, 80 grammes.
> Pure Distilled Water, 1 litre.
> Iodide of Potassium, 25 grammes.

with the subsequent addition of a two *per cent.* barium nitrate solution; this and the following formulæ are taken from Abney's excellent treatise, to which the reader is referred for fuller information on the subject.

The developing solutions are made as follows :—

> Ferrous Sulphate, 40 grammes.
> Glacial Acetic Acid, 30 to 40 c.c.
> Alcohol, 2 s.
> Water, 1 litre.

The addition of a very minute quantity of gelatine (dissolved in water), to this developer, adds to the density and fineness of the deposit a very desirable result in Micro-photography.

The amount of alcohol in the solution depends on the age and condition of the bath; with a new bath none is required.

The iron should be dissolved before the other ingredients are added.

When the picture lacks density, the following intensifying solution will answer.

> Pyrogallic Acid, 4 grammes.
> Citric Acid, 4 to 8 grammes.
> Water Distilled, 1 litre.

to this a few drops of

> Silver Nitrate, 2 grammes.
> Distilled Water, 50 c.c.

must be added immediately before its application to the negative.

After the plate has undergone treatment with these solutions, the bromides and iodides unaffected by light are dissolved away, by immersing in

> Hyposulphite of Soda, $\bar{3}$ i.
> Water, $\bar{3}$ vi.

After washing and drying, the plate may be varnished with any of the usual negative varnishes sold by photographers.

The nitrate of silver bath should be exposed to the light, strong sunlight if possible, when not in use.

When required, it is filtered, and a few crystals added until it regains its proper strength. An argentometer is very useful for ascertaining the amount of silver in solution.

We now proceed to the manipulations necessary for the preparation of the sensitive surface, before exposure in the camera.

A plate being removed from the dish, we clean it as described, taking care to avoid touching the surface with the fingers, then either grasping the right hand upper corner with the thumb and finger of the right hand, or supporting the plate with a plate holder, we pour the collodion on it, forming a pool at the same corner, then flowing it to the opposite, then to the left hand bottom corner, and finally after completely covering the plate, off at the right hand bottom corner. It being now held in a nearly vertical position over the mouth of the collodion bottle, a gentle rocking motion is imparted to it to prevent its setting in ridges. When no longer tacky, which is known by barely touching one of the bottom corners with the finger, it is placed in the nitrate of silver bath, by a steady downward motion of the dipper, moved about for a few seconds, and finally left for a period varying from one to five seconds, according to the temperature and condition of the solutions.

It is then removed, and (after all superfluous fluid has been drained off) placed in the dark slide, ready for exposure in the camera. This should take place as soon as possible, as all delay tends to injure the plate.

The dark slide enveloped in a dark cloth is now placed in the camera, care being taken to keep the side of the plate which was uppermost in the bath, in the same position in both dark slide and camera.

The requisite exposure being given, the slide is carried back to the dark room, its original position being still maintained The developer is poured into a glass measure, and the plate covered with the solution in the same manner as it was with collodion, care being taken that none flows over the edges.

If the picture flashes out at once, over-exposure is indicated; the developer should immediately be dashed off, and the plate washed under the rose tap.

Should it develope gradually the developer should be swept slowly backward and forward over the plate, as long as the action is apparent, and removed before any symptoms of fogging make their appearance.

The plate is now thoroughly washed, and if not sufficiently dense, the intensifier is poured over it, while upon it a few drops of nitrate of silver is placed in the measure; the intensifier is then poured back, and the solution again poured over the plate. Intensification requires great judgment and should only be continued as long as all detail is perfectly clear; the instant, or better before, the slightest fogging appears, the intensifier should be removed and the plate again well washed.

We now immerse it in the hypo-solution, where it should remain till all the unaltered salts are dissolved away, this is known by the creaminess having disappeared from the back of the plate; it is then washed in frequent changes of water for several hours, and finally put aside to drain.

When dry, the plate is warmed before a clear fire, the temperature not being raised higher than the hand can very comfortably bear, the varnish is then applied in exactly the same manner as collodion, and when all superfluous fluid has drained away, heat is again applied; the temperature being carried much higher than at first, to prevent the formation of a mat surface. Should the intensifier not be used and the plate found to lack density after fixing, the mercurial solution mentioned in the following process may be applied, (as little exposure to actinic light as possible being given before its application).

When the reproduction of very minute markings is necessary, dry plates never give as good results as the wet, at least in our hands; they are, however, sufficiently perfect for all practical purposes, while the shortness of the exposure, cleanliness of manipulation, and when purchased, the time saved by the absence of any previous preparation, almost counterbalance this disadvantage.

Great care is required in using these plates, as the slightest glimpse of actinic light will utterly spoil them. A double thickness of ruby glass must be substituted for the ordinary yellow glass of the dark room, and even of this light, as little as possible should be allowed to reach them. The dark slide should be carefully examined, and if of the double dry form, two sheets of zinc the same size as the plate, each covered with black velvet on one side and having a spring between them, should be placed with the velvet side next the back of the plates, in the dark slide.

These plates can now be purchased so cheaply, with the degree of sensitiveness as regards collodion marked on each box, that we do not recommend the amateur to attempt their manufacture. As a practical knowledge of their preparation may be an assistance, a description is appended of a process which has afforded excellent results.

The preparation of the following solutions may be carried out in daylight.

> No I. Bromide of Potassium, 43 grs.
> Nelson's Gelatine, 10 grs.
> Distilled Water, 6 drachms.
> One *p. c.* solution of hydrochloric acid, 10 minims.

This solution, is put aside, and we place in another earthenware vessel—

> No. II. 60 grains of special hard dry plate gelatine, covering it with three or four ounces of water.

Both are allowed to soak for twenty-four hours.

All the following manipulations are done in non-actinic light.

> No. III. Nitrate of Silver pure, 50 grs.
> Distilled Water, 1 ounce.

Nos. I. and III. are placed in hot water till their temperature reaches 120° Fahr. No. I. is now gradually added to no. III with constant stirring and shaking, no. III being placed in a bottle for the purpose; this process should take at least half an hour.

No. IV. Iodide of Potassium, 4 grs.
Distilled Water, 2 drachms.

is now added, well shaken, and the whole boiled for three quarters to one hour or more, according to the sensitiveness required. No. II. from which all the water has been removed being gradually added with constant stirring during the boiling.

The emulsion while hot is poured into a bowl to set, and when cold placed in three thicknesses of coarse muslin, through the pores of which it is wrung out into a deep earthenware vessel. It is washed with six changes of water with frequent stirring for one hour. The ringing and washing being repeated, the emulsion is thrown upon a hair sieve to drain, a process that will take at least an hour.

We again dissolve it by heat, and after filtration through cambric, neutralize it by the addition of a few drops of a dilute solution of ammonia, (if great sensitiveness is required).

Two drachms of alcohol are finally added and the whole put away for a couple of days.

The most difficult parts of the process are now before us, and the amateur will find his skill, in-

genuity and patience severely taxed, but practice
will enable him soon to overcome the difficulties.

Having cleaned and prepared the plates in the
same way as when about to coat them with collodion, we hold them in exactly the same manner,
that is, between the finger and thumb of the right
hand, or by a pneumatic holder. One is now warmed
and a little more than one drachm of the melted
emulsion poured on, spreading it evenly over the
surface with a swaying motion. When covered,
keeping it horizontal we place it on some perfectly
level surface, (such as a large sheet of plate glass),
to set.

A sufficient number being prepared and allowed
to set, they are placed in a drying closet. Of these
there are innumerable forms, the arrangement we
leave to the reader, simply stating that the plates
must be kept horizontal and as strong a current of
heated air as can be obtained, allowed to pass across
their surfaces. Every glimpse of actinic light being
excluded during the process.

For further details we recommend Abney's Emulsion Processes.

The plates being prepared according to the foregoing instructions, or better still procured from
some respectable firm, the following articles are obtained.

A dish, preferably of ebonite, just large enough to
hold a quarter plate.

Another of the same material to hold four or a
larger number will be found very useful, as a great

saving both of time and solution is effected if several plates, which have each received a similar exposure, are developed at the same time.

A dish of earthenware to hold a saturated solution of alum, another for a similar solution of mercuric chloride, and one for hyposulphite of soda or fixing solution:—these should all differ in appearance, to avoid the possibility of putting a solution into one, which had previously contained another.

The dishes being placed in a convenient position in the dark room, one or two plates according to circumstances are placed in the dark slide; every precaution being taken during this and all subsequent operations for the exclusion of actinic light. The exposure being made, the slide with its contained plates may be put aside, till another batch has been exposed, or for an indefinite period, and developed at leisure.

The following formulæ are recommended by Mr. Swan for the development of the dry plates prepared by him.

There are two different developers, alkaline pyrogallic acid and ferrous oxalate. The former is preferred by many, especially for views and portraits, on account of the control it gives the photographer over the character of the resulting negative. We shall describe it first, although the latter has given more satisfactory results in Micro-photography.

No. I. Liquor Ammonia, (·880) ʒ ii.

Bromide of Ammonia, ʒ 1½.

Distilled Water, ℥ 20.

This solution will keep for some time if the bottle containing it is securely corked. Care should be taken that the ammonia is the full strength specified, if weaker the whole formula will be ruined.

No. II. Pyrogallic Acid, 4 grs.

Water, 2 oz.

This solution should be mixed *immediately* before development.

The plate or plates being removed from their respective slides, are placed in the ebonite dish.

An equal quantity of no. I. solution is added to no. II., and the whole poured equally over the plates which are left at rest in it.

As it would be difficult to weigh a less quantity of pyrogallic acid, and as the resulting solution is far too large a quantity for a single quarter plate, we recommend the development of at least four at the same time.

Should the picture flash out suddenly after the application of the developer, over exposure is indicated; the developer should be immediately poured back into the measure, the plates washed for a second or two under the tap, and left covered with water in the dish while one or two drops of a solution,

Bromide of Ammonia, $\mathfrak{Z}$j.

Water, $\mathfrak{Z}$j.

are added to the developer.

The water is poured from the plates, the developer again applied, the process carefully watched, and the solution poured off before any symptoms of fogging appears in the whites of the negative.

In the opposite case, where the image is slow in making its appearance, the solution should be poured back into the measure, and a quantity of no. I. solution, varying from ½ to double the quantity at first used, added to the developer, and at once applied to the plate, without previous washing.

It will be seen that owing to the variations admissible in the strength and preparations of the solutions employed, the condition and density of the negative is almost completely under control, if the exposure has been kept within anything like reasonable limits.

The developer is now poured off and the plate washed for a few minutes. If not sufficiently dense (especially when owing to under exposure), the washing is continued for some time, and the ordinary pyrogallic acid and silver solution, mentioned under the wet process, employed.

Whether intensified or not at this stage, the negative is quickly washed under the tap and placed in a solution, consisting of—

Hyposulphite of Soda, ℨ ii.
Water, ℨ x.

this solution should be fresh.

The plate is left in it for about five minutes after

all creaminess has disappeared from the back of the film, and then washed in frequent changes of water for not less than six hours. It is then placed in a saturated solution of alum for fifteen minutes, and again thoroughly washed.

After this washing, if the plate on examination appears to lack density and be deficient in vigour owing to over-exposure, it is immersed in a saturated solution of bichloride of mercury, until perfectly white; again thoroughly washed for several hours, and finally placed in a solution consisting of—

Liquor Ammonia, ℨ iii.
Water, ℥ x.

which changes the white colour to a dense black.

By modifying the strength of this solution, any density between that produced by the mercuric chloride alone, and its complete reduction to the metallic state by the ammonia, may be obtained, if the plate is washed immediately on the appearance of the desired colour.

The plate is put aside to dry spontaneously, as the application of heat is liable to melt the film and destroy the negative.

Varnishing is unnecessary.

If only one plate is developed at a time, or the method of development described, considered too complicated, Edwards' Developer may be used.

Two stock solutions are prepared as follows :—

No. I. Pyrogallic Acid, ℥ i.
 Glycerine, ℥ i.
 Methylated Alcohol, ℥ vi.

Mix the glycerine and alcohol, then add the acid.

No. II. Bromide of Ammonia, gr. 7½.
 Liquor Ammonia ·880, ℥ i
 Glycerine, ℥ i.
 Distilled Water, ℥ vi.

Add one part of each to fifteen parts of water and use in equal proportions.

These solutions will keep a long time if well corked.

Ferrous oxalate or Eder's developer on account of its simplicity and cleanliness, has, lately become a favourite with amateurs. It possesses few other advantages except that it can be used side by side with the wet process, an impossibility where ammonia is given off, as it contaminates the atmosphere of the dark room.

The following solutions are used :—

No. I. Neutral Oxalate of Potash, ℥ vi.
 Boiling Water, ℥ xx.

Place the oxalate of potash in an earthenware vessel and add the *boiling* water, stirring with a glass rod till sufficiently cool to be poured into the bottle intended for its reception. If on testing with red litmus paper the solution is found to be alkaline, sufficient dilute sulphuric acid is added to cause a very faint acid reaction with blue litmus paper. If

too acid, carbonate of potash may be added to nutral-
ize the excess. The solution is now securely corked·

> No. II. Sulphate of Iron, ℥ iii.
> Distilled Water, ℥ vi.

This solution is also securely corked.

Suppose we wish to develope a quarter plate;
four parts or ℥ viii of no. I. solution are placed in a
glass measure, and one part or ℥ ii of no. II. solution are added; not *vice versa*, or a precipitate of
oxalate of iron is formed which weakens the mixture. One or two drops, according to circumstances,
of the bromide of ammonia solution, the same as
that used as a restrainer in alkaline, pyro developer
is added to the whole. The plate is placed in the
vulcanite dish, the developer caused to wave to and
fro over its surface, till the required density is obtained; the same precautions as those previously
mentioned with regard to fogging, etc., being observed.

After development the mixed solution is transferred to a separate bottle holding about eight or
ten ounces; when full, this is boiled in an iron sauce-
pan for ten minutes, returned to the bottle and
securely corked; it will answer admirably for the
development of several future plates or paper pictures, if always boiled after use, and the clear liquor
poured off. ·

Fixing and subsequent intensification are proceeded with in exactly the same way as for the
other processes.

The blackening of the plate with ammonia, after whitening with mercury, is carried on in the open air, at all events in another apartment, as it would spoil all subsequent manipulation with the wet plate process.

It has been recommended to add citric acid to the alum solution when pyro is used and oxalic acid with ferrous oxalate, to remove any stain due to the developer on the resulting negative. There should be no stain after development, and we cannot too strongly condemn the use of acids in any of the solutions used subsequent to development. Without entering into the theoretical grounds of this advice, we simply state that nothing tends more, except traces of hypo, to impair the permanency of the negative.

CHAPTER VI.

ARRANGEMENT OF APPARATUS, ETC.

MOST methods adopted in these pages are taken
from the works of others, at least the principle is
the same although the arrangements are slightly
modified. Our claim is simply to have shown the
possibility of obtaining results with an ordinary
parafin lamp, equal if not superior to those hitherto
obtained with sunlight; thus placing Micro-photography within the power of all.

It would be an impertinence as well as an injus-
tice to those from whom we have learned much,
if no mention was made of their work in these
pages. We have no intention of attempting a
classification on the ground of superiority or priority
but will mention what we know, in the order it
happens to present itself.

All microscopists are acquainted with the beautiful photographs of N. Rhomboides, published by
Mr. Woodward in the Journal of the Microscopical
Society. The arrangement he adopted was very
similar to that employed with the eight immersion
by us, and with which the photograph of S. Gemma
was taken. He seems to have always used sunlight
and the wet process, which as a rule give results far
superior to those obtained by lamplight and dry
plates.

Few amateurs can ever produce anything equal to his: first, because every assistance was given by a generous government, who placed all instruments likely to aid in the production of perfect results at his disposal: secondly, a peculiar training is required for the attainment of that skill, of which only a few are capable, but which must be acquired before any progress can be made in a special art, when it arrives at that perfection which brings it just within the borderland of science.

We can therefore hardly blame those who, perhaps more enlightened than ourselves, perceive, that even if we had the opportunities of Mr. Woodward, we lack the ability to use them.

Mr. Wenham, as early as 1855, gave instructions in the *Quarterly Journal of Microscopical Science*, for improving Microscopic Photography. We do not recommend the plan *then* adopted. Mr. Highley we believe subsequently obtained very good results, but his method rendered necessary the use of a special apparatus, distinct from the ordinary microscope.

Captain Abney has given us a few good hints, and we regret he has not told us more. The impression left, after a perusal of his opinions, was that he must have worked with inferior lenses. He considers that higher powers than the quarter inch do not give as good results, on account of "the diffraction images of parts of the object . . . . . this being dependent on the relative sizes of the object and of the aperture of the objective." This is directly contrary to our experience, as no objectives we possess give as good results as the eighth and one twenty-fifth of Messrs. Powell and Leland. This we think was owing to the proper relation between the size of the object and the aperture of the glass being strictly observed; the adjustment for cover glass being made at the same time with extreme accuracy; and I think principally because the balance between the power and aperture of the condenser, and that of the object, was more easily equalised with these objectives than the others.

Good results can never be obtained unless we are careful to employ suitable objectives with certain objects. Low powers should not possess excessive angular aperture, as their utility depends more upon their penetration, (that is their power of showing with sharp definition different parts at different depths), than it does on their resolution, (or capability of differentiating very fine peculiarities of structure lying on the same plane). On the other hand high powers are not required to show the relation existing between the different parts of a structure lying at different depths, but to define the delicate markings on the surface of those, whose relations at different depths have already been investigated by lower powers. What has before been said about the relation existing between NA and penetration, will show the folly of expecting a lens to combine, great penetration, magnifying power, and angular aperture, at the same time.

Lately M. Tisandier has described three methods adopted in Paris. All have at least one great disadvantage; they require the employment of electricity in its most expensive form, this places them beyond the reach of ordinary individuals.

In the first the virtual image is transformed into a real one by displacing the eyepiece, so that the image, formed by the objective, falls on the opposite side of the ocular; this may answer with the one inch and lower powers, but with higher powers the light is so much diminished by the interposition of four surfaces of glass between the objective and image, that focussing becomes impossible.

The second method due to M. Vogel, consists in placing a short focus photo lens, as well as the ocular, between the objective and image. This method is evidently worse than the first, as we have eight fresh surfaces interposed. It reminds us of a person trying to do two things at the same time, which are better done separately.

In the third and last a considerable deviation from the others is noticeable. The ocular being removed, a short focus carte lens is substituted and the whole apparatus is placed in the horizontal position. This improves the illumination as the mirror is not required to reflect light on the object. The same objections, however, hold good with regard to this. But if a sufficiently strong light can be obtained, we are confident that results exceeding those of our Parisian brethren are possible, as according to M. Tisandier the highest magnification was eight hundred diametres.

It has been the custom with some microscopists

to place a concave, and we believe sometimes a convex lens, between the objective and focussing screen, when increased magnification is required. Mr. Woodward was the first to suggest this practice; it is not to be recommended, as the increased magnification is more easily obtained by enlarging the negative; besides any combination above the objective impairs the image. It is our firm conviction that nothing has been such an obstacle to the practice of Micro-photography, as the failures consequent on allowing the eyepiece to remain in its usual position.

We shall now point out a few of the advantages of our own method.

Any microscope can be fitted with the necessary additions at a trifling cost, the amount depending on what can be done by the possessor himself; in fact, the majority of first-class instruments, already possess these additions.

The same apparatus answers with every method of illumination, it can be made to assume any position, the use of oblique light is facilitated, and the employment of photography with objects in fluid becomes possible.

On account of its portability the several changes are easily made, while the simplicity of its construction prevents the possibility of its getting out of order.

In the following pages we have adopted the plan of describing a different method of illumination, and arrangement of apparatus with each objective, in order to illustrate the employment of different sources of light; not on account of any advantage gained by using the illumination and arrangement described with that particular objective. For instance, sun light, lamp light, and electricity, give equally good results with the eighth immersion; if the adjustments of the microscope are suitable for this objective, and the conditions of their employment identical.

With regard to illumination, we most decidedly recommend the use of an ordinary parafin lamp, with one broad single wick, one and a quarter inch or more in width, in preference to all other kinds; it is always obtainable, requires little care, and the expense incurred is very trifling.

If properly employed a negative can be taken by this light in five minutes, with the one twenty-fifth, having a magnification from 1000 to 5000 diameters, which will bear subsequent enlargement to 50000 before the finest details become visible to the naked eye.

Swan's incandescent lamp has numerous advantages, among which may be mentioned the facility of focussing, on account of the whiteness and brilliancy of the light, and the possibility of placing the apparatus in any position during its employment.

Care should be taken that the carbon filament occupies the centre of curvature of the glass globe, otherwise aberrations will be produced by the inclination of the surfaces; this might be avoided by partially flattening the globe on two sides, one being behind and another in front of the edge of the carbon filament.

The only serious objections to the use of Swan's lamp for Micro-photography, is the care and attention required when the electrical supply is produced by a battery. The enormous expense entailed by the purchase and working of electro-motors, placing them beyond the reach of private individuals, in fact, until electricity is supplied by accumulators or from a central station parafin will be found far superior.

The actinic power of the electric light is inferior to the intensity of its illumination, allowance must be made for this when judging of the time necessary for correct exposure.

Grove's or Bunsen's elements are the best form of battery; bichromate cells from their inconstancy being a continual source of trouble.

The lamp should be covered and mounted in the same way as the parafin lamp before described; or one may be substituted for the other, if a flat hook capable of slipping into an arrangement on the pinion block of the stand, is fastened to the upright back of both.

Magnesium in the form of Solomon's lamp may be employed with low powers, when it is necessary to shorten the exposure; but with high powers want of light when focussing is the difficulty to be contended with. The costliness of magnesium renders its prolonged use inapplicable to this purpose.

Lime light gives excellent results, but the danger and trouble connected with its use, along with the impossibility of moving it about and centering, prevent its extensive employment.

A minute intensely brilliant spot of light, requiring no care, free from danger, and capable of being
placed in any position, is required by those who cannot afford steam power and electro-motors. Up to
the present nothing fulfils these requirements as well
as a parafin lamp with a bull's eye condenser and from
the short strides electrical illumination has made
within the last twenty years, no substitute is likely
to make its appearance from that quarter.

Those resident in cities have a great advantage
over their country neighbours, (who require photography more, on account of their isolated position),
in possessing gas, from which we may now expect
something, owing to the stimulus given to its improvement by the introduction of electricity.

In every practical art the rule has been to proceed
from simpler methods to those more difficult. We
shall not depart from it, but endeavour to select such
examples as will lead gradually to a full understand-
ing of the theoretical and practical principles involved,
the increasing difficulty of manipulation at the same
time producing the requisite amount of skill.

It is almost impossible to plan a method in which
the practical manipulation and theoretical knowledge
proceed hand in hand; we therefore recommend a
perusal of the following chapters, before attempting
to photograph by any of the methods described in this.

A note book is a necessity; every proceeding
should be entered, to enable us to repeat any particular plan previously adopted, and from a com-

H

parison of failures and successes to eliminate the causes producing the former.

The following is an example of the form of entry adopted by us:

Apparatus Horizontal.

I. Object, coscinodiscu from guano, mounted in balsam and bensol equal parts.

II. Objective, quarter inch, corrected $10°$ for cover glass.

III. Condenser, two French combinations, together about half an inch in focal length, and distant $\frac{9}{16}$ of an inch from the object.

IV. Aperture of diaphragm quarter of an inch in diameter.

V. Lamp 13 inches from stage, edge of flame towards the object.

VI. Bull's eye condenser four inches focal length, four inches from flame of lamp with convex side turned towards object.

VII. Ground glass of camera twenty-four inches from object.

VIII. Focus till the dots appear surrounded by a red areola.

IX. Sulphate of copper cell placed midway between bull's eye and substage condenser.

X. Gelatine dry plate 10 times collodion.

XI. Exposure one minute.

XII. Ferrous oxalate developer, with one drop of bromide solution to the ounce.

FIG. 24.

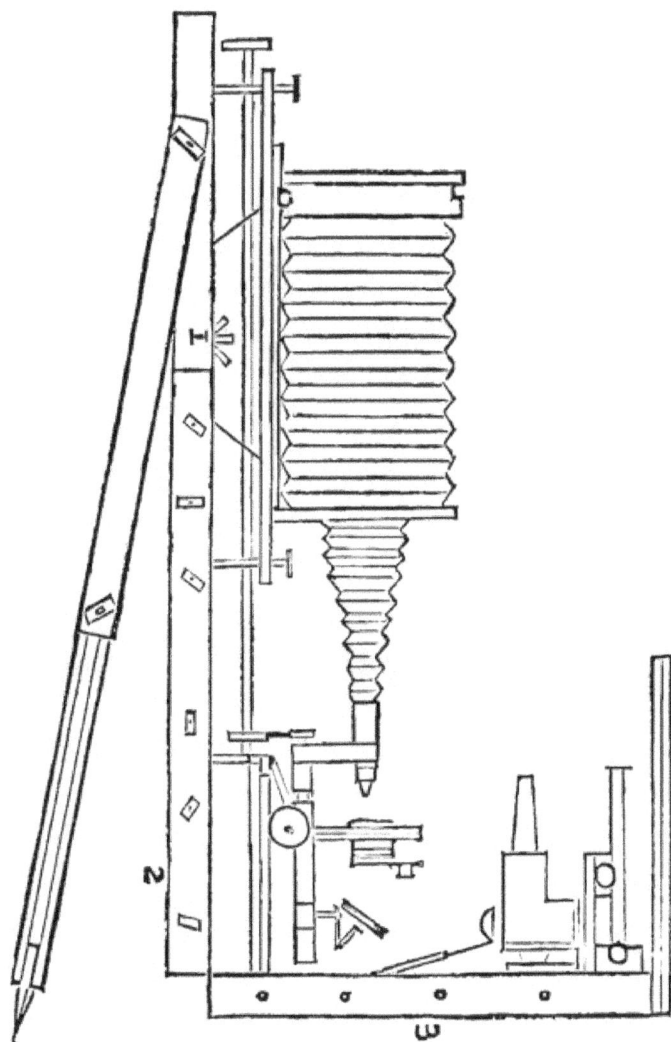

XIII. Result, completely successful, or over or under-exposed as case may be, in this case very much over-exposed.

It will be seen from the foregoing notes the necessity for having the laths at the side, marked off in inches and half inches.

We now proceed to photograph with the one inch objective. The object is supposed to be mounted in fluid, as an illustration of the method to be adopted under these circumstances.

The microscope being placed on the platform, the divisions, nos. 1 and 2, of the stand are rigidly fastened together, by tightly screwing down the board fitting on the bolts projecting from their sides.

They are then raised till at right angles to no. 3. The camera legs are fastened to the end bolt on no. 1, their pointed extremities resting on the ground, maintain this end of the stand in the perpendicular position.

The horizontal position of the stage is ascertained by a spirit level.

The tube containing the eyepiece, if not previously attached to the microscope, is now put on; the objective we wish to use, in this case the one inch, is adapted.

The substage apparatus, *with its condenser and pin hole cap*, is fixed in position.

The lamp previously trimmed and lighted, is placed on no. 3 division of the stand, about ten or twelve inches from the mirror below the condenser.

The largest hole of the diaphragm is turned till it occupies a position approximately in the optic axis of the microscope. The plane side of the mirror is brought as close as possible to the stage of the microscope, without interfering with the motions of either itself or the condenser.

The lamp should now be brought near the stand, on which it rests, (*i.e.*, as low as possible), by the rack and pinion movement attached to its back.

Our reason for keeping the lamp low and the mirror high, is, the greater the angle made by the incident and reflected rays, the better the illumination.

The light from the lamp may be thrown on the condenser by altering the inclination of the mirror, the eye at the same time being applied to the eye-piece.

Still looking through the microscope, the pin hole cap is brought into focus, and centred, by turning the screws in the rim of the substage tube.

When central, the diaphragm plate is turned till a medium-sized hole is under the condenser; this aperture is now brought into focus by increasing the distance of the objective from it, and made coincident with the optic axis of the instrument, by displacing the diaphragm wheel sufficiently in its own plane.

This being accomplished the flame of the lamp is focussed, by altering the distance between the top of the condenser and the objective, (in other words by using the coarse or fine adjustments of the micro-

scope), and centered by altering the inclination of
the mirror. We now make the image of the flame
of the lamp itself, and that of its reflection in the
mirror placed behind it, coincident. This may be
done by giving a slight twist or tilt, as the case may
be, to the tin cover surrounding the lamp. The
bull's eye condenser is placed from ½ to 2 inches
in front of and from the flame of the lamp, with
its convex side facing the microscope, and centred by observing the position of its image in the
microscope; this will be known by its appearing as
a brilliant disc of light without flaw or spot, in fact
a regular sun, in the centre of the field of vision.

We now remove the condenser from its tube; this
is best done by racking back the objective, and un-
screwing the lenses from the diaphragm tube at the
object side of the stage, without disturbing the
diaphragm or any apparatus except those requiring
removal. If on account of their position they cannot be removed in this manner; the substage tube
II, is unscrewed without altering the position of
either the diaphragm or mirror. The condenser can
then be easily taken out and the substage tube replaced, the same precautions being observed as for
its removal.

All the apparatus required are now perfectly central with each other. It might be thought that
keeping the condenser in the substage would be
superfluous, but by its aid the images of the diaphragm, lamp, and bull's eye condenser are observed through the microscope, and all **made**

coincident with its optic axis. Perfect accuracy of adjustment in this respect, is one of the greatest aids to the production of good and truthful negatives.

The object, chosen by the rules laid down in the following pages, is placed on the stage and brought into focus. If the light is too strong, the distance of the diaphragm is increased, or a smaller aperture substituted, taking care that it occupies exactly the same position with regard to the optic axis of the apparatus as the former one, and *vice versa*. It should be born in mind that the less light incident on the object, consistent with clear definition and subsequent facility in focussing, the better.

It would be much simpler to centre the tube holding different sized discs, as they might be substituted for each other without any danger of altering the position of the other parts, but withdrawing the inner tube would very likely displace the mirror; this plan is more suitable when the apparatus is placed in the horizontal position, no mirror being then necessary. Whichever plan is adopted, the differentiation of the structure should be distinct; the markings dark; their interspaces bright and clear. The attainment of the proper means will be facilitated by remembering that too great a quantity of light impairs definition; too little increases the difficulty of focussing and unnecessarily prolongs the exposure. Intensity is what is required, not quantity.

The best light being obtained, the eyepiece tube of

the microscope is removed, and the camera arranged vertically over the microscope, in the following manner.

The camera is fixed over the microscope by the binding screw attached to the stand beneath it, the conical bellows being removed from its front. The round disc of light formed on the ground glass is placed exactly central by means of the levelling screws at each corner of the oblong board, to which the camera is attached when this is accomplished, the binding screw is tightened, and the conical bellows slipped into its original position on the front. of the camera. Great care is now required in attaching the end of the conical bellows to the short tube of the microscope. It will be much more easily done if the interior of the tube fixed to the end of the conical bellows is covered with black velvet, as well as its interior; it will then slip easily into that of the microscope, and the entrance of all light rays will be prevented.

The image of the object is now roughly focussed on the ground glass; the desired magnification obtained by racking the ground glass backwards or forwards to increase or diminish it, bearing in mind that the actinic power of the light is said to decrease in proportion to the square of the distance from its source.

The part of the object we wish to focus should, if possible, completely fill the plate; this is sometimes impossible, as its thickness prevents the use of lenses of sufficient power to magnify the lateral dimensions, while their penetration defines those parts lying on different planes.

The most satisfactory magnification being obtained, we pass the silk cord round a bevelled ring attached to the fine adjustment of the microscope, and the wheel on the rod passing under the camera stand. The focussing glass previously adjusted, is applied to that part of the image on the ground glass, requiring the most accurate definition. The best focus is obtained by observing the appearance of the structure through the focussing glass during the revolution of the milled head attached to the rod connected with the fine adjustment.

The chemical focus is found when the object appears clearly defined, although its markings are surrounded by a red areola, (see *previous pages*). A black cloth is now hung over camera and stand, but not reaching lower than the stage of the microscope; the blue glass or sulphate of copper cell placed between the bull's eye condenser and the mirror; and an opaque card between the two latter, to enable the exposure to be given at the proper time.

We now proceed to the dark room, and having lit the lamp, observe the condition of the trays and measures, their position, and that of the various bottles containing solutions.

The barrel should contain a good supply of fresh and pure water; the tin box containing the dry plates is placed on the floor or a shelf, on the opposite side of the dark room, to that occupied by the washing trough and lamp.

We recommend the use of dry plates twice as fast as ordinary collodion in this case, as the in-

tensity of the light is insufficient and the position of
the camera unsuitable for wet plates.

The dark slide is examined; the side on which the
plates are inserted opened, and placed with the zinc
and velvet partitions, on some *dry* support.

The door of the dark room being securely fastened
and the curtain drawn across it, the tin cover is
placed over the lamp.

The tin box containing the packages of dry plates
is now opened; the lid should fall back to a horizontal position, thus forming a shelf on which the
pasteboard box and cover can be placed during the
removal of the desired number from it. As we are
now using a double dry slide, two are, removed,
and held one above the other by their edges between the thumb and finger of the left hand, with
the right we replace the cover on the cardboard box
containing the remaining plates, shut the tin box,
and if necessary, cover it with a dark cloth kept for
the purpose.

Holding the plates in our left hand, we go to the
support on which the dark slide has been placed,
and insert one plate with the gelatine surface downwards or looking towards the unopened side. The
gelatinised surface may be distinguished from the
other, by its want of lustre, when reflecting the
image of the lamp; care being taken to use only
sufficient light to enable us to barely distinguish the
difference. One zinc partition is placed with its
velvet side next the back of the plate, the other
with it in the opposite direction. The second plate

is laid with its back on the velvet side of the second
zinc partition, its gelatinised surface then looks towards us. Whatever plan is now adopted for fastening the plates in position, great care must be
taken to avoid touching them with the fingers, during the process; as should this happen an indelible
stain will make its appearance at that spot during
the subsequent development. Having fastened
them we shut up the dark slide or plate carrier,
and enveloping it in a dark cloth, or better placing
it in the inside breast pocket of our coat, we carry
it to the apartment where the microscope and apparatus are arranged.

The right hand side of the cloth hanging over the
apparatus is slightly raised, the dark slide slipped
quietly and carefully underneath it, and substituted
for the ground glass of the camera.

The sliding door covering the plate *near the microscope* is then drawn out, and a delay of a few seconds
made to allow all vibration to cease.

The utmost care is required for these manipulations, and if the dark slide and camera are not accurately made, some sticking or jarring very likely
to cause displacement and endanger the centering
of the apparatus is certain to occur.

During this interval a watch is placed on the stand
near the lamp where the time can be easily observed.
The opaque card is now gently removed, the requisite exposure, say ten seconds given, and the card
gently replaced; the dark slide shut, the carrier removed from the camera, and placed with the ex-

posed plate next the person in the breast pocket of the coat; every precaution being taken as before, to prevent displacement.

Now if uncertain as to the proper amount of exposure required, or desirous of photographing another portion of the object; we remove the card, re-examine the image, and correct the focus if necessary. The card being replaced, the dark slide is again substituted for the ground glass, care being taken that the plate farthest from our person while the slide was in the pocket, is placed facing the microscope; the sliding door is withdrawn; a slight rest allowed, and the requisite exposure given as before, the same precautions being observed.

Having exposed the required number of plates, we carry them to the dark room, and either put them carefully away till a more convenient time, or proceed at once to develope by one of the processes suitable for dry plates; preference being given to the ferrous oxalate, on account of its giving perfect results, where the time of exposure, fastness of the plate, and intensity of the light are previously known.

With the same arrangement, it is quite possible to photograph opaque objects, and those illuminated by dark ground or oblique illumination. Wenham's parabolic illuminator, a spot lens, or what is the same thing, a central stop in the condenser, are all used for the latter purpose. We, however, prefer the following simple arrangement.

A piece of electro-plated copper, bent in the form

of a ring B, fig. 25, is substituted for the tube fitting
into the substage aperture of the microscope. About

FIG. 25.

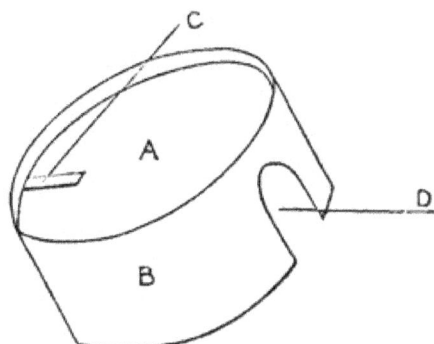

an eighth of an inch from its upper edge, we place a
cardboard disc with a slit C in it; all the interior of
the tube is blackened except a narrow strip, corresponding in size and position to the slit. This part
of the ring should be flattened. An opening D is
made in the front, to enable the angle at which the
light falls on the object to be changed. The ring B
is covered on its outside with black cloth, so that it
may slip easily up and down in the aperture, in the
centre of the stage.

The light is concentrated by the bull's eye on the
narrow strip corresponding to the slit, and from
thence on the object.

The next method is the only one possible, when it
is desired to photograph objects in fluid, with arti-

Fig. 26.

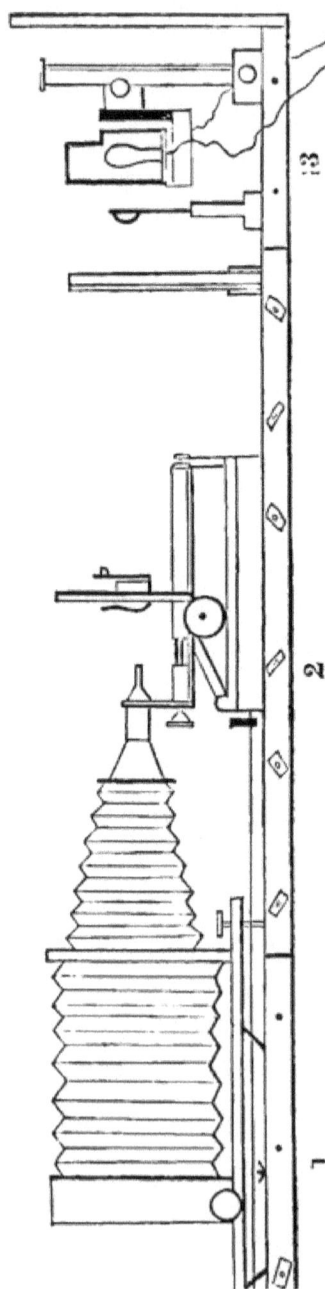

ficial light, and which require high powers for their resolution. We shall confine ourselves to the $\frac{1}{4}$ inch objective. The apparatus is considered as still remaining in the position adopted in our first experiment.

The lamp stand, and bull's eye condenser, are clamped in their original position by binding screws fitting in the division running down the centre of plank no. 3.

The camera legs are shortened to allow of the upright division no. 1 and no. 2 assuming the horizontal position, but need not be removed (they are omitted in the figure). The side board is detached, and replaced two bolts lower down. When firmly screwed on, it keeps all three divisions of the plank perfectly rigid, and in a straight line with each other.

The camera and sulphate of copper cell are removed; Swan's incandescent lamp substituted for the parafin, the edges of the carbon filament being turned towards the microscope.

Connection is now made with the source from which the electricity is supplied. We shall here only mention that electro motors are preferable to a battery, but unfortunately the great expenditure necessary for their maintenance, as well as first purchase, puts their application to photographic purposes beyond the reach of ordinary individuals. In a late number of the *Photographic News* an electric generator worked by hand is proposed, which may prove practicable if sufficiently low in price.

If battery power is used, thirty cells of Groves' battery will be found sufficient for a full sized Swan's lamp We have had no experience of the smaller forms recently made, but from a consideration of the character of the light required, we have no hesitation in recommending them.

One or two cautions are necessary with regard to the batteries, and lamp. Bichromate cells are useless for many reasons but principally for not being sufficiently constant. Some means of regulating the current is useful, as the sudden passage of a strong current is liable to cause the lamp to burst. The ordinary method of attaching the lamp by a spring is most objectionable, as the slightest movement on the same floor or adjoining apartment, will cause sufficient tremor to utterly ruin the resulting negative.

The whole stand is now raised to the perpendicular; the microscope stage accurately levelled, and supported in that position by lengthening the camera legs.

The eyepiece tube being fitted to the microscope, a one inch or lower power is screwed on, and the condenser, with its pin hole cap, lamp, and bull's eye centred as before. We remove the inner tube, or central sliding diaphragm as it is called, from the cloth lined tube fastened to the largest hole on the wheel diaphragm. The latter being approximately centered under the condenser, we insert the inner tube fitted with a perforated disc, (the perforation being about one-tenth of an inch in diameter or less) till it touches the surface of the lowest combination of the condenser. It is now made accurately central with the pin hole cap, by bringing it into focus, and displacing the wheel diaphragm sufficiently. The low power objective is now racked back, and the quarter inch substituted.

The pin hole cap is removed and the perforated disc brought into focus; if not exactly central it is made so by the centering screws fitted in the rim of the diaphragm tube.

At this stage in the arrangement of the apparatus it is well to consider a peculiar method of illumination first described by Dr. Dallinger in the *Microscopical Journal*, and I believe called by him the sun light. The following description will we hope enable others to obtain it. Unfortunately so much is still left to individual intuitive skill, owing to no fixed rules being found by which we might ascertain the relative proportions of diaphragm aperture, power and NA of condenser, intensity of illumination, and other factors, that its accomplishment will at first be found difficult.

The lamp and bull's eye being centered, the perforated disc is brought into focus with the quarter inch objective and also centred. This must be done by the centering screws, to prevent disturbance of the relative positions of the substage apparatus. The aperture in the disc, when the sunlight is obtained, should appear as an intensely bright spot in the centre of an illuminated field A, fig. 27.

The chances are a hundred to one that it appears as represented at either B or C. If as at C it shows that either the lamp or bull's eye condenser is entirely out of the optic axis of the instrument; if as at B, one or other is too much to the left of the

Fig. 27.

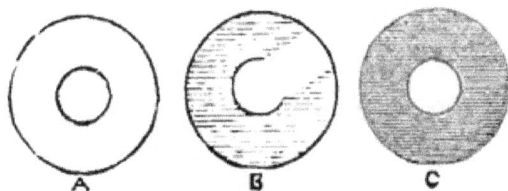

instrument; the condenser and diaphragm are supposed to be already accurately centered, and in fact must be if the foregoing directions have been properly carried out.

Whichever appearance presents itself the bull's eye condenser is turned to one side, and the image of the lamp brought into focus. If the flame is not central it is made so, by the rack and pinion, or lateral screw motion attached to its stand. The

I

position of its reflected image in the mirror behind it should be observed, and if not coincident with the flame itself, it is made so by tilting the tin shade surrounding the lamp. The bull's eye is again brought into view and centered. If the sun-light is not yet perfect or if it appears as at B, the lamp is moved more to the left or the right according to the position of the bright ray proceeding from the aperture of the diaphragm. These manipulations are greatly facilitated by remembering the relation the bright ray bears to the position of the lamp and bull's eye; for, when considered as pointing from the periphery to the central hole, it indicates directly the position of the lamp and bull's eye with regard to the optic axis of the instrument. At the same time the distance of the bull's eye from the lamp may be slightly altered, care being taken to keep it central. When all other means have failed, a very slight alteration of the position of the condenser by the centering screws, will sometimes cause its almost magical appearance. For further details we refer the reader to Dr. Dallinger's communication.

Having obtained the sunlight by some means or another, we rack back the objective. Now if the object is flat with fine markings on or near its surface we place it on the stage, but should it consist of delicate scarcely differentiated structure, lying on different planes, we replace the pin hole cap, which, by the by, should itself be accurately centred, and if not, rejected, and another tried. *With the objective corrected for uncovered*, (the ultimate correction for thickness of cover glass being made according to the method described on page 17, chap. ii.) we bring the object into focus. If the light is too great or deficient, the condenser is slightly moved farther from, or nearer to the object, or the inner tube containing the perforated disc may be removed and another substituted, and the tube returned to exactly the same position, care being taken to avoid disturbing the relative positions of any of the other apparatus. During its slow removal the eye should be kept on the object, as frequently a slight alteration of the distance of the perforated disc from the glass slip, especially when the light is too great, produces the desired effect. With regard to the proper distance of the condenser from the object; we have found double its focal length, with the bull's eye one inch from the lamp flame, the most favourable position as a rule. Much, however, depends on the thickness of the glass slip, on which the object is mounted. Some recommend the image of the lamp focussed on the object, this practice is only mentioned to be condemned.

Having once found the proper distance for a condenser with any particular object, the number of divisions marked on the tube of the substage apparatus, if entered in our note book, enables us to repeat it *ad infinitum*.

The proper quantity, and intensity, of illumination obtained, we proceed to place the camera in position, observing the precautions mentioned when describing a similar process with the one inch.

The actinic focus is found in the same way, and the camera being covered with a dark cloth, the sulphate of copper cell is clamped in position.

The light is sufficiently intense for the wet plate process. We therefore proceed to the dark room, and sensitise a plate according to the instructions contained in a previous chapter, drain it as quickly and thoroughly as possible, and place it face downward in the dark slide used for that purpose.

If it was necessary to caution the amateur against touching the sensitised surface with his fingers in the dry process, ten times the precautions are necessary here, as the slightest touch will infallibly ruin the plate.

We shut the slide and convey the plate in as horizontal a position as possible to the camera, care being taken to avoid its knocking against anything or being shaken during the journey. After insertion it is exposed in the manner previously described, three to six minutes being sufficiently long, and when removed from the camera immediately carried to the dark room.

Any development suitable for the wet plate process, is at once proceeded with. The resulting negative, after fixing and subsequent washing, being dried and varnished.

During the description of this method exception might be taken to needless repetition. We have only one excuse to offer. Having seen numerous failures and the destruction of many plates caused by the neglect of some trifling detail only remedied by perseverance, we prefer repetition on our own part, to failure on that of our readers.

The following arrangement should be selected by those compelled to adopt one plan alone. Its advantages are, the ease and quickness with which the different manipulations are performed; the small number of adjustments necessary; and the absence of all danger of disturbing the centering of the different instruments, as the microscope and camera are not placed in direct continuity. In the former arrangement this was very liable to occur, when fixing the bellows front to the camera and microscope, or when substituting the dark slide for the focussing screen.

Perhaps its greatest advantages are the possibility of focussing at any distance from the microscope and of manipulating without leaving the dark room.

The drawback connected with this method, is the impossibility of photographing objects in fluid, on account of the horizontal position of the microscope. For this reason also we should be extremely careful to accurately arrange the distance of the condenser from the slide, because if the heat rays are brought to a focus on the object, even when an ordinary parafin lamp is the source of illumination, those mounted in balsam are liable to change their position, owing to the melting of the medium during exposure.

The recess jutting into the dark room, full description of which has been already given, being constructed, the entrance of all actinic rays is provented, by pasting several folds of yellow or black paper, (the latter being preferable) over both interior and exterior. The edges of the sleeves where they join the partition are further secured by tacking thin deal laths round them, which are also covered with black paper. A black cloth curtain with an orifice for the tube of the microscope and fine adjustment, should be placed inside the dark room so that it can be let down when the arrangement of the apparatus is complete.

We suppose the apparatus continuing in the position assumed during the manipulation necessary for the process last described.

The stand is placed in the horizontal position by the removal of the camera legs; all the instruments are detached from it and carefully put aside; the side boards are changed from their original position and shifted two bolts lower down, or till flush with the extremity of division No. 3, and firmly screwed on. This keeps the two last divisions perfectly rigid, and enables us to fold division No. 1 under them; where it also is firmly secured by one or more of the binding screws, running in the slits with which the centres of Nos. 1 and 3 are furnished.

Thus folded up and tightly secured, the stand is placed, with the platform for the microscope in front, on the stand forming the floor of the recess.

The ruby lamp is now lit in the dark room, and all necessaries (such as developing solutions, dry plates, dishes, etc.) examined and put in their proper places.

FIG. 28.

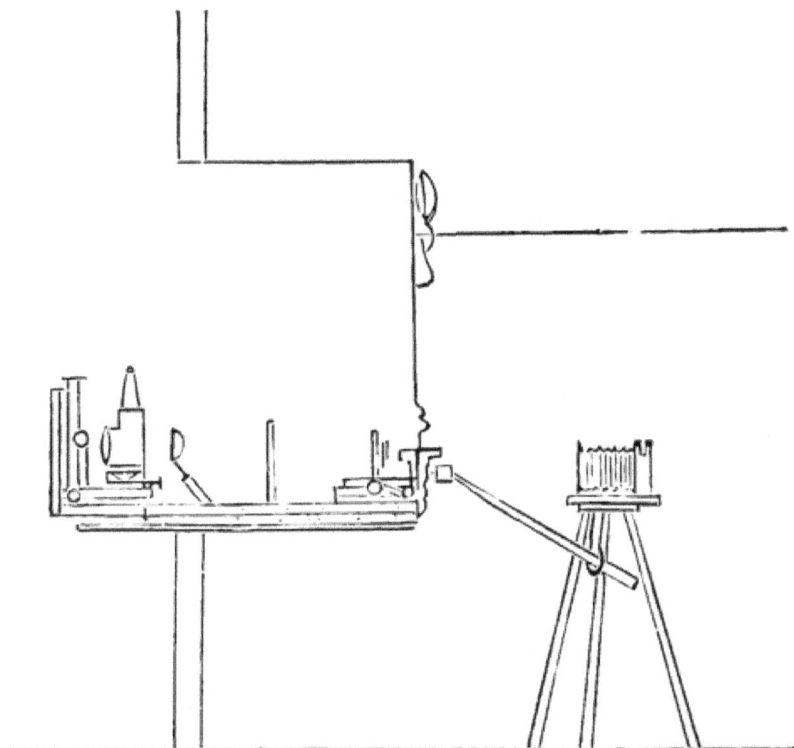

The microscope lamp or Swan's electric light, the former being preferred with the power, we are about to use, is got ready and placed on the stand. The microscope with eye-piece attached being fixed in position, the centering is carried out by a low power objective as before described. All these manipulations are performed by the hands through the sleeves, or better through the windows in the same position. These windows should of course be capable of subsequent closure by sliding doors. The sleeves were only mentioned on account of being easily made, and their completely excluding actinic light.

Whichever method is adopted, before centering with the low power objective, we may substitute the dry front of the $\frac{1}{8}$, if it possesses one, for the top combination and pin-hole cap of the French condenser. It can be easily made to fit it; if too large, by a card-board ring slipping on the screw of the French lens; or if too small by placing it in a cardboard tube fitting round the outside of the two French lenses.

This arrangement may appear subversive of the rules already laid down with regard to the achromatism of the condenser, as the front is supposed to consist of a single uncorrected plano convex crown. Practically this objection does not hold good, as the light is monochromatic. Spherical aberration is produced but the errors from it are counter-balanced by the increase of NA. which now bears almost exactly the proper proportion to that of the objective itself. It will be seen that the definition is perfect, and the ease with which the sunlight illumination is obtained is remarkable. The low power objective being attached to the microscope, we proceed in the same manner with the centering as in the last method.

The object is placed on the stage of the microscope: it should be remembered that when using high power such as the one-eighth imm. it is better to select a flat thin object, as the angular aperture and magnifying power are too great to allow of a large amount of penetration. The part we wish to photograph being chosen, the low power is removed.

The ¼ objective, if a water immersion, may be wetted with either water or glycerine. The former is better as far as definition is concerned, but the latter on account of not evaporating, allows of a greater length of time for arrangement of apparatus, and a more protracted exposure.

The proper focus and correction for cover glass obtained, the eye-piece tube is removed, and the sleeve securely fastened round the short tube, by a strong elastic band in addition to that already inserted in its orifice.

Before proceeding with this description, it should be mentioned, that the most favourable position for the condenser is about one and half to twice its focal length from the object, and the sliding diaphragm is undoubtedly the best to use with this and all higher powers. The best sized hole is that which, when brought into focus, is slightly smaller than the field.

The fine adjustment wheel is screwed off, and the screw passed through a small hole in the sleeves. A light rod attached to a cork, as before described, is fitted to the wheel; this is now returned to its original position on the microscope.

The camera, from which the oblong board and conical bellows front have been removed, is attached to an ordinary camera stand, and placed so that an image of the object may be properly received on the focussing screen.

The distance of the screen, owing to its not being continuous with the microscope, is only limited by the length of the dark room. Practically it will be found that perfect results are not obtained when it exceeds three feet. It should be also remembered that the actinic power of the light decreases as the square of the distance of the plate from the source of illumination, and the duration of the exposure varies inversely as the actinic power.

Seating ourselves behind the camera and taking the focussing rod, (which may be conveniently supported within reach by a string) in our left hand, we apply the focussing glass to the screen with the right. When the proper focus is obtained the rod is let gently down. The sulphate of copper cell or blue glass is placed in position; the windows closed or sleeves fastened; and the curtain which has an orifice for the tube of the microscope only, allowed to envelope the whole of the recess, except the tube.

Two tightly stretched strings or laths, reaching for about four or five feet beyond the camera and about seven from the floor, will be found very convenient to hang a black cloth over, covering the camera and making it practically continuous with the microscope during exposure. The possibility of any actinic light, except that from the microscope, reaching the sensitive plate is thus prevented.

A cap is placed over the orifice of the microscope and we proceed to place the dry plate in the carrier. Twenty times collodion should be used. Great care must be taken to avoid shake and prevent the entrance of actinic light. The subsequent manipulations are identical with those already described.

The following notes of the manipulations employed for the production of the Negative, from which the print of the Bacillus Anthacis in the frontispiece was produced, may prove of use as an illustration.

Apparatus, horizontal, placed in recess of dark room, camera, not attached to microscope.

1. Section of lung containing Bacilli Anthacis prepared by Ehrlich's method.

2. Objective one-eighth imm., corrected to thirty-five for cover glass.

3. Condenser single French front, with dry eight top, one sixteenth inch from object. Aperture of sliding diaphragm as close to the back of condenser as possible, and covering half the diameter of the field when brought into focus with one-fourth objective.

4. Blue glass near condenser.

5. Bull's-eye eleven inches from stage of microscope.

6. Lamp one inch from bulls eye paraffin and edge of wick.

7. Ground glass of camera twenty-four inches from object.

8. Usual focuss.

9. Glycerine and water the immersion fluid.

10. Gelatine plate twenty times collodion.

11. Exposure three minutes.

12. Ferrous oxalate developer; two drops of fifty p. c. bromide solution added to one ounce.

13. Result shown in photo.

The last method requiring description is that in which sunlight is used.

Any of the previous arrangements may be employed if the apparatus is situated so as to allow the light from the heliostat to fall on the condenser.

It will be found advisable when the distance between the microscope and reflecting mirror exceeds six feet, to place a plane convex lens, (whose focal length is about equal to a quarter or a fifth of the distance, and whose diameter is not less than one twelfth its focal length) with its plain side towards the microscope, so that an image of the sun may be formed about two inches from the diaphragm wheel. This is easily done by placing a card at the diaphragm wheel to receive the image of the sun.

The stand, from which the lamp and bull's eye condenser are removed, is fixed in the horizontal position, on a firm support, so that the second mirror of the heliostat may be easily made to occupy a a position coincident with the optic axis of the microscope.

The heliostat should be arranged and left working for a quarter of an hour previous to the arrangement of the apparatus for photography. This enables us to test the accuracy of its adjustments, and allows the action of the first wheel full time to be communicated to that on the mirror.

These arrangements made, the microscope is fixed in position, the sulphate of copper cell placed between it and the heliostat and everything carried out in exactly the same manner as described when photographing with Swan's incandescent lamp.

The sixteenth immersion being employed, the dry front of the eight will be found to answer admirably when used as in last process.

All subsequent operations are conducted in exactly the same manner.

Dry plates or the wet process may be used, the former if ten times as rapid as collodion requiring an exposure of from one to five seconds. No fixed rule can be given for exposure with sun-light for the reasons already mentioned.

In conclusion we must caution our readers against expecting success at the first trial, unless they have had some previous experience in photography. On this account we recommend the production of negatives with an ordinary camera and lens before attempting micro-photography.

When carrying out the methods here described, avoid speaking or walking in the same or adjoining apartment during exposure as this is very likely to cause vibration of the instruments.

Too great care cannot be taken to prevent the entrance of actinic light. All instruments must be lined with black velvet. When in position the microscope and Camera should be covered with a dark cloth before the substitution of the plate carrier for the ground glass and subsequent exposure. Every means must be adopted to prevent the entrance of actinic light into the dark room, and while placing the sensitised plate in the carrier as well as during developement, no more light should come through the ruby glass, or exposure of the plate to it be allowed, than shall just enable us to work.

Before concluding our remarks on this part of of the subject, we would impress upon the photographer the necessity for determining the magnifying power in every case. This may be done as follows:

After removal of the carrier containing the sensitised plate from the camera, an ordinary micrometer slide divided into one-hundreths and one-thousandths of an inch is substituted for the object. Its image is focussed on the ground glass, and the division measured with small rule. For example if one of the divisions corresponding to one-hundreth of an inch on the micrometer measures two inches on the ground glass, the magnification equals two hundred diameters.

It is not necessary to adopt this method in every case, for, if the magnifying power of a certain objective is known when the image formed by it is at two separate distances from any object, the magnifying power can be determined when used at any other distance.

Suppose the magnifying power with the quarter inch six inches from the object, has been found to be fifty diameters, and at two feet and a half, to be two hundred diameters. If on a subsequent occasion we are photographing with the same lens, and the screw is twenty-four inches from the object, the magnifying power would be one hundred and fifty diameters.

It is sometimes difficult to tell if the sun illuminates the object equally when a heliostat is not

used. The most convenient method of ascertaining
this, is by observing the image formed on the sensitised plate through a piece of ruby glass let into the
shutter of the plate carrier, or in the top of the
camera itself. The former is the better method with
dry plates, the latter with wet.

CHAPTER VII.

DEFECTS IN NEGATIVES, THEIR CAUSES AND CURE.

Before attempting the production of negatives with the microscope, we recommend the beginner to obtain some skill in ordinary photography, by the perusal of a standard work on the subject, with the subsequent reproduction of views by the camera. This will at least prevent the disappointment occasioned by unsatisfactory results, arising from want of skill or errors in development. It should, however, be remembered that a wider and more scientific knowledge is required than is generally obtainable by these means.

The ordinary photographer has only to place his camera in such a position, that a picture formed by artistic rules is thrown upon the screen. Having given an exposure suitable to the intensity of the illumination, fastness of the plate, and tints in the picture, a latent image is obtained which on the application of a well chosen developer, reproduces every characteristic detail in the original view.

The Micro-photographer in addition to a knowledge of the same kind, must recognise and allow for the chemical changes produced in different sensitive compounds by light; which is itself modified by agencies beyond the perception of unaided vision.

Many of these changes and agencies have only

lately been discovered. A very little experience will convince us, that others exist, which, owing to a want of continuity between cause and effect, are incapable of demonstration, in the present state of our knowledge.

Some are certainly dimly present to our minds, and although not fully recognised, are allowed for by an unconscious cerebration, if we may be allowed the term.

The best method of arriving at a knowledge of how to produce good negatives, is by studying the defects of bad ones, endeavouring at the same time to discover their cause, and the remedies for them.

Inaccuracy in the preparation of solutions and want of manipulative skill, are the principal difficulties besetting a beginner. It will only be necessary to mention where and how these inaccuracies are most liable to occur, to enable the manipulator to guard against them.

When distilled water is mentioned never use any other; pure spring water, not rain, being employed for its production. Neutral solutions must be tested, and what are sold as neutral salts can never be taken upon trust, but must be made so. Obtain pure chemicals; see that all solutions are of the proper strength; keep all bottles requiring it tightly corked; and filter solutions when directed. Have the dark room well ventilated, and on no account allow a trace of ammonia to be present where the wet plate process is employed.

Among errors of manipulation may be classed

K

those in which the plate is fogged by the entrance of actinic light. This is most liable to occur when working with very sensitive dry plates. An enumeration of the various means by which it may gain access to them and the precautions to be taken for its prevention, although incidentally mentioned in other places, will be repeated here.

The dark room may allow light to enter by the door, or the ruby glass window. The former may be remedied by hanging a curtain both outside and inside, the latter by the addition of another sheet of ruby glass or non-actinic paper. The lamp may be defective, and is almost certain to be so unless covered by the tin shade, with a box beneath it, as before described.

The box in which the plates are stored may not be light tight; keeping it enveloped in a rug or dark cloth is the best preventative, care being taken that the dark room door is properly fastened before the box is opened. When a plate has been removed the box should be shut and covered up at once, otherwise it might easily be forgotten.

Prolonged exposure of the plate to the light, even when filtered through two thicknesses of double flashed ruby glass, will produce a fog. The interval therefore during its transmission from the box to the plate carrier should be as short as possible.

Every point of the plate carrier must be examined, and its accurate closure ascertained, before the plate is consigned to it; and afterwards covered with a dark cloth.

The camera is to be examined in the same way. Entrance of light need not be apprehended, if it and the microscope tube are lined with black velvet, and when in position covered with a dark cloth. The substitution of the carrier for the focussing screen should take place underneath this or another cloth, the rays from the lamp being cut off from the microscope by a blackened card placed between them, before the drawing up of the slide and its replacement.

Actinic light must be guarded against after removal from the camera, and during the subsequent operations; the slightest glimpse will produce a fog on the negative. Fortunately this may be removed by immersing the negative in a solution of bichromate of potash, care must be taken that the clearing off is not carried too far else weakness may be produced. Another kind of fog may occur when using wet plates, owing to alkalinity of the bath. A similar condition of the oxalate of potash solution produces a whitish deposit on the film of dry plates, it is therefore advisable to add a little sulphuric acid to this solution after its complete neutralisation.

Pin-holes are due to dust on the plate. A soft large camel's hair brush specially kept for the purpose should always be passed over dry plates before placing them in the dark slide. Black markings are generally caused by the plate coming in contact with the fingers, or some unclean surface, and insufficient drainage before placing it in the carrier.

Frilling is very liable to occur in hot weather with

gelatino bromide plates. It commences by a blistering of the film at the edges, and its extension may be prevented by immediately immersing the plate in a saturated solution of alum. This also serves to clear the negative.

Blurring of the image may be due to a variety of causes. The principal are tremor of the apparatus during exposure, and reflection from the back of the plate. A black velvet backing similar to that already described will entirely prevent the occurrence of the latter.

Reflection from the surfaces of the crystals suspended in the film has been mentioned, and dying the film suggested as a remedy. This lessens the rapidity too much to be available in micro-photography.

It is well known that the invisible part of the spectrum acts on bromo-iodised plates, and from the results obtained with mono-chromatic light, we are inclined to refer a blurred condition of the image to the unequal refrangibility of the different parts of the spectrum. It is for this reason also that the employment of cobalt blue glass increases the distinctness of the image.

The character of the negative differs considerably from that required in ordinary photography. All beginners as a rule fall into the error of supposing that the image by the object must be opaque.

convey a cheap idea of the rarer of microscopic objects by engravings. All micro-photographs should be transparent in the high lights, because there are parts of certain objects, the relative brightness of which may easily escape notice in the microscope owing to their minuteness. For example: the visibility of the structure of the diatomaceæ depends on the formation of extremely bright spots or lines on a ground, itself possessing considerable brightness, while the surrounding field has an intensity less than the former and greater than the latter. Now if the field be opaque the finer details of the original structure will be lost, not only by a want of variety in the different shades, but also by a blocking up of the minute spaces with a dense deposit.

Great density is preferable to weakness. The former may be easily remedied by flooding the plate with a weak solution of perchloride of iron, and when the required transparency is obtained, washing the negative rapidly in flowing water.

A weak image is produced by so many causes, that a complete enumeration would be impossible without entering farther into the subject than is required for our purpose.

The principal are however:

I. A weak bath solution, or too strong a developer with wet plates.

II. Over exposure, or unsuitable condition of developer with dry plates.

III. High refractive index of the medium in which the object is mounted. This prevents proper differentiation of structure both visually and chemically. Even when it does not do so visually it seems to disturb the corrections of the object glass, so that a weak negative is the result.

IV. Non-actinic colour of medium. This and the former fault, are especially noticeable with balsam. Diffusion of the colouring matter of the object, particularly if red, through the medium may also weaken the image.

V. Delicate structure of the object produces a weak image. This may be obviated by staining the specimen with carmine, which cuts off most of the actinic rays, and if cobalt blue glass is placed between the source of illumination and the object, it stops off the remainder. The resulting negative will be more distinct than the visual one.

VI. Specimens stained with logwood or blue pigments are unsuitable for micro-photography.

Objects of a yellow colour or great delicacy give the best results with wet plates. Those of a red colour or coarse structure the best with dry.

With different kinds of development a plate will give different results. For instance, where it is necessary to reproduce a great variety of shades, alkaline pyro development is superior to ferrous oxalate, with dry plates. No intensification should be attempted with wet plates, other than that obtained by ordinary development. This is especially the case, when the negative is required for enlargement.

For direct printing, a thin negative obtained with a wet plate, sometimes gives better gradations when subsequently intensified with silver. A dry plate, if the ferrous oxalate development is stopped before the slightest fogging appears, generally requires intensification with mercury and ammonium hydrate after fixing, especially if intended for the production of lantern transparencies by contact.

From experiments with different sensitive compounds it seems probable, that iodide of silver, as its maximum intensity ends suddenly in the indigo portion of the spectrum, will give the greatest differentiation, and should therefore be used with delicate structures.

Bromo iodide emulsions, depend for the chaacter of their sensitiveness on the condition to which the emulsion is brought as regards colour during the cooking process, and as this sensitiveness extends considerably into the invisible part of the spectrum, a determination of when and with what structures they should be used, becomes a very difficult matter, and beyond the scope of a work of this description. We shall therefore content ourselves with the indications for their employment already given.

The amount of intensification required will be found to vary directly as the magnifying power of the objective.

To sum up. The kind of plate to be employed, depends on the character of the object, nature and intensity of the illumination, and magnifying power of objective. The development depends on the purpose for which the resulting negative is intended, the character of the object, and the kind of plate.

It is often impossible to obtain a sufficiently dense

negative directly. When this is the case, the weak negative should be copied in the enlarging apparatus (to be described), using a very weak illumination. A positive will be obtained which on development will be at all events as dense as the negative, and when intensified much more so. If then another copy be taken from this positive, a negative is obtained, very much denser than the original negative.

This simple enumeration will be sufficient to enable any intelligent person to avoid the defects mentioned, or to apply the proper remedies for them. For fuller imformation we refer our readers to any of the numerous text books on this subject.

CHAPTER VIII.

THE PRODUCTION OF POSITIVES ETC.

Silver printing by contact calls for little comment, as the process is much simplified by the sale of sensitised albuminised paper possessing the great advantage of keeping for many months without deterioration.

In addition to a few dishes, a printing frame is the only apparatus necessary. It consists of a shallow wooden box, the bottom of which is formed of a sheet of thick plate glass. The negative is placed on the glass, a sheet of sensitised paper over its prepared side, then a thick cloth pad, and over all a hinged back. Two cross bars having springs attached to them, are hinged to one side of the box, and when pressed down on the hinged back, can be fastened in position by clips fixed on the opposite side.

The process may be watched by unfastening one of the cross bars and lifting that side of the sensitised paper; when released it will fall back to its original position on the negative. The character of the resulting picture can be modified to a great extent by a proper consideration of the kind of light required, and very artistic effects may be produced by shading under-exposed portions of the negative. An intense diffused light will be required for a dense negative, while one possessing the qualities before recommended will require a subdued light and two or three hours exposure for the differentiation of the markings and the proper reproduction of the tones existing in the original. Bright sunlight must never be employed for silver printing by contact.

The depth of the print requires some practice to determine. As a rule the printing must be carried farther than will be necessary in the finished picture, because it loses intensity in the subsequent toning and fixing processes. These processes should not be commenced until several prints have been taken, as one print uses as much solution and gives as much trouble as a dozen.

When a sufficient number have been printed, they are placed in a large quantity of water in a darkened room. The water is changed every hour or so until all milkiness disappears, and they are then immersed in a solution consisting of—

> Gold Trichloride, gr. iv.
> Sodium Acetate, gr. cviii.
> Distilled water, $\mathfrak{Z}$ x.

This solution is best made with boiling distilled water if immediately required for use. We prefer to make it the day before.

A large flat dish will be found the most convenient for holding the prints and solution, as they require to be kept in continual motion during the process. The tint must be deeper, and as in printing, the process must be carried farther than is necessary in the finished picture. A very deep chocolate brown verging on purple gives the most artistic appearance when finished. Should an engraving tone be required the process must be continued until a decided purple tint is produced. Overtoning is to be avoided as a cold grey picture is the result.

After toning the washings are repeated and the prints finally fixed in a solution containing—

Sodium Hyposluphite, ℥ is.

Liq. Ammon. Fort. ℳ iii.

Water, ℥ ix.

This solution should not be above 85° or below 75° Far. and always freshly made.

After ten or fifteen minutes immersion, with continual motion, the unaltered silver will be dissolved away. The prints are then immediately removed to a large tub of water, and kept in continual motion for a quarter of an hour. The water is then changed; at the end of an hour the change repeated; at two hours again repeated, and so on for eight hours. At the expiration of this time all traces of hypo will have disappeared; the prints are then hung up to dry.

All these operations, except the final washings must be conducted in a subdued light, coloured light will not answer, white light alone enabling us to distinguish the proper tint when toning. The prints when dry are cut to the required shape, with a sharp knife, not scissors, and mounted on cardboard with thick starch or glue, to which a few drops of carbolic acid have been added. When perfectly dry their appearance may be much improved, by passing through a rolling press, or in its absence pressing them with a heavy hot smoothing iron, a sheet of clean paper being interposed between the surface of the iron and the print.

We strongly recommend the employment of a professional photographer for the production of these prints from ordinary negatives, as no amateur can finish them off so well, without a useless expenditure of time and money.

An excellent method of producing glass or paper positives by development, has lately been introduced to the notice of photographers. It is more suitable for amateurs than the process just described, as the tedious washings between the application of the different solutions is avoided, and the development itself reduced to the simplest form by the use of the ferrous oxalate process, employed for dry plates. The results obtained by the employment of Swan's opal plates, leave nothing to be desired as far as artistic display and ornamental work is concerned. Many of the most beautiful prints of polycystinæ and diatoms especially coscinodisci we have ever seen, were produced by this process.

The simplicity of the manipulations and the direct production of a positive by one process, is itself a recommendation to those who do not require a multiplication of proofs. All that is necessary after exposure, development, and fixing in the ordinary manner, is the application of a saturated solution of bichloride of mercury, until the film becomes perfectly white. After the plate has been well washed and allowed to dry spontaneously, a black backing is attached. This produces that exquisite transparency and softness so seldom seen since the Daguerreotype process became extinct.

For ordinary scientific work, however, the paper is preferable, not only on account of its superior distinctness, but also because storage is facilitated by its lightness, and its examination, by the absence of any risk of fracture when handling it.

Should it be desired to produce this paper at home, it is only necessary to coat ordinary albumenised paper with the emulsion described on a previous page, as follows :—

The melted emulsion is poured into a flat ebonite tray, slightly larger than the sheet of paper to be coated. The sheet being held at two opposite corners, so as to form a kind of angular spout, its most central and lowest part is allowed to rest on the emulsion, the ends at the same time being allowed to sink gradually until the whole sheet rests evenly on the surface of the emulsion. This gradual descent prevents the possibility of any air bubbles remaining beneath the sheet. Should this however happen, the end nearest the bubbles is raised and again allowed to sink gradually, when it will be found to push them before it in its descent. The sheet is allowed to rest for a period varying from one to ten minutes, according to the temperature of the emulsion and finally hung up in the dark to dry. If carefully guarded from actinic light and damp, this sensitised paper will keep indefinitely.

All the trouble of the foregoing manipulations may be avoided, by purchasing the paper ready coated. An excellent quality is supplied by Morgan and Co. In fact we recommend all to whom time is of value, to purchase everything from those whose business it is to manufacture articles of this description, as in the majority of cases home production is false economy.

The extreme sensitiveness of this paper renders printing by contact, especially by daylight, a rather difficult matter, unless the light is subdued by placing tissue paper or muffed glass a couple of inches in front of the printing frame. The exposure can then be prolonged, a great advantage when accurate results are required. It should be remembered that the intensity of the illumination and duration of the exposure have more influence on the condition of the resulting positive than the character of the negative itself. For instance, a very thin negative may be so weakly illuminated that a harsh positive is the result; while a very dense negative, if exposed to bright sun-light, will produce a positive completely wanting in proper contrast. Advantage should be taken of this and the illumination arranged to suit the character of the negative.

It is evident from this that no rule can be given for exposure when daylight is employed, and even with lamp light, owing to the variations in density of different negatives, an approximation only is possible. At the same time an over-exposed negative always gives the best results, as a weak light and under-exposure is all that is necessary for the production of a good positive. We have found fifteen seconds, three feet from an ordinary paraffin lamp, sufficient exposure for a negative possessing the characteristics recommended in a former chapter. When no other means of printing but by contact are available, a denser negative would give better results.

Copying with the camera is to be preferred to direct printing; first, because the exposure is lengthened; and secondly, because the character of the resulting picture depends more on the size of the stop, and intensity of the illumination than on the density of the negative, and as a micro-negative is less dense than an ordinary negative, greater power of modifying the positive is placed in the hands of the operator. Any form of doublet may be used. The ordinary Petzval portrait combination is to be preferred. It should be capable of taking a cabinet picture.

The ordinary quarter plate negative is the most convenient to copy from, especially for enlarging, because the pictures produced by a combination covering that sized plate, are more perfect in detail, and when artificial light is employed, it is possible to obtain good illumination without the use of very large condensers or equally large photographic combinations, which in addition to their costliness possess inherent faults that cannot be corrected.

If we desire a copy the same size or smaller than the original; the brass tube containing the lenses is unscrewed from the camera; the cell containing the back combination removed; the lenses are taken out of it; the ring separating them is next removed and put back in the cell, and the lenses replaced in their original order but in contact with each other. By this means the definition is considerably improved, and the consequent decrease of focal depth prevents the images of objects, situated beyond the film of the negative, being impressed on the sensitive surface.

The modified objective is fitted to the camera and levelled at the negative. The latter should be placed in a window having a northern aspect and an uninterrupted view of the sky. To obtain this it is sometimes necessary to tilt the camera upwards, great care is then required to keep the surface of the ground glass parallel with the negative, and both perpendicular to the optic axis of the objective.

The most accurate focus, with the full aperture of the objective is obtained; a stop of sufficient size introduced, and the camera covered with a black cloth. A great improvement will be effected, and the entrance of all extraneous light prevented, by pinning the black cloth round the negative, so as to form a sleeve between it and the camera; it must however be left unfastened at the bottom to allow of the removal of the cap on exposure.

Before describing the remaining manipulations it is necessary to remind the operator, that the aperture of all stops or diaphragms are best named in terms expressing their relation to the focal length of the objective. As $\frac{f}{10}$ when the aperture is half an inch and the focal length of tho objective ten inches and so on, the smallest aperture generally employed being one fiftieth of the focal length or $\frac{f}{50}$. A combination of great focal length will bear a smaller aperture proportionally than a shorter one. For instance, with a negative possessing the characteristics before mentioned, a quarter plate photographic objective requires an aperture $= \frac{f}{25}$ while one $= \frac{f}{35}$ will be necessary when using a half plate portrait combination.

We now proceed to the dark room, and having cut our gelatino-bromide paper to the required size, float the unprepared side on the surface of some perfectly clean water, (the greatest care must be taken to prevent the water from flowing over the prepared side). When perfectly flat it is laid evenly on a sheet of patent plate and finally placed in the dark slide. Some recommend the complete immersion of the paper in water before laying it on the glass plate, but tho length of time necessary for perfect drainage, and the risk of spoiling the picture if sufficient time is not allowed for it, has led us to adopt the former plan, which, if it requires more skill and care, has at least the advantage of celerity.

After an exposure of say ten seconds, we return to the dark room and on removal of the paper from the dark slide, replace it in the water. A sufficient quantity of ferrous oxalate developer, in the proportion of five parts of a saturated solution of oxalate of potash to one of ferrous sulphate (care being taken to add the latter to the former not *vice versa*, else a precipitate is formed which weakens the developer), and one or two drops of Bromide of Ammonium are placed in a glass measure. The mixed solutions are then poured over the paper, previously placed in a developing tray after its removal from the water. We may now utilise all the ruby light at our command to watch the process of development, as the colour of the solution covering the paper is sufficiently non-actinic to prevent the image being injured by too strong a light. In about half a minute the picture will begin to appear and in about three minutes the process will be complete. It must not be carried too far, in fact not so far as would be desirable in the finished picture, as the intensity is considerably increased by the fixing solution. After a slight rinse in water it is immersed in the alum bath for one minute, and after another rinse placed in the hyposulphite solution for three minutes. It is then washed in frequent changes of water and finally dried and mounted in the manner described when treating of ordinary silver prints, with this exception, that the picture when pasted on cardboard must be allowed to dry perfectly before the application of the hot iron, or the gelatinised surface will be certain to adhere to anything placed over it during the process. We wish here again to impress on the operator the necessity of using the developing solution perfectly fresh, as the formation of the ferric salt from use and exposure tends to slow the process with consequent weakening of the image, to say nothing whatever of its destructive effect on the silver compounds which form the image.

These positives may be intensified by mercury and ammonium hydrate in the same way as ordinary dry plates, but we do not recommend this method, as the prints are liable to turn yellow on the application of a hot iron.

Enlarging. There are innumerable methods employed for this purpose, we select two as being most suitable for our work, and requiring very little additional apparatus.

The first method is identical with that last described, except that the lenses composing the objective are reversed, *i.e.* the front lenses are placed in the cell containing the back conbination, with the convex surface towards the screen, while the back combination is placed in the cell, from which the front lenses have been removed, so that the lens which formerly faced the screen, now faces the object to be enlarged. If the cells themselves will fit either end of the tube, this reversal is easily accomplished without removing the lenses. It should always be remembered, that the lenses forming what is considered as the back conbination in ordinary photograhy, must be placed in close apposition, and the ring which separates them placed in the cell before them.

Every part of the negative not required for reproduction should be masked, and all extraneous light cut off.

After focussing, the smallest stop compatible with a proper differentiation of the several shades in the negative should be inserted, and an exposure according to the size of the enlargement and density of the negative being given, the usual manipulations are proceeded with.

The advantage this method possesses over that to be described, is the possibility of enlarging from a negative of any size. This, however, can be accomplished by other means, and is only mentioned as an additional resource where time is important, or where the markings are so fine that they would be injured by the coarseness of the film on dry plates, when the collodion process is not available for the production of a reduced negative.

An ordinary magic lantern, the condensers of which are not less than four and a half inches in diameter, answers excellently when placed in the recess of the dark room. The entrance of actinic light being prevented by fastening the sleeve intended for the microscope over the nozzle of the lantern.

The focus is obtained by pinning a sheet of white paper on a moveable stand, and substituting the sensitized surface for it after placing a cap over the nozzle. The latter operation is much facilitated, if a piece of ruby glass is let into the cap.

Many artists prefer the delicate tone of the ordinary silver print to that produced on a gelatino-bromide surface. All that is then necessary is the production of a positive on paper, if a negative by daylight; or on glass, if one by lamplight is required. This method has the advantage of allowing the production of any degree of density in the several proofs by subsequent intensification.

By varying the processes, immense power is given the operator, and no matter what the character and condition of a micro-photographic negative, it is quite possible to change it completely without altering the essentials of the structure depicted on it, thus rendering possible the production of a print from what would otherwise be considered a useless negative.

It might be objected that the whole process, from the first production of a negative with the microscope to the finished print, was too tedious to be practicable.

This is by no means the case, because the production of a negative only takes one hour, and several can be produced together with a very little increased expenditure of time. The subsequent washing is not considered, as it requires no special attention. The printing by camera, contact, or enlarging, takes about half an hour, and the finishing about another half hour, therefore from beginning to end the actual time consumed is about two hours.

APPENDIX.

PRESERVATIVE MEDIA.—The most generally useful medium for mounting microscopic objects is Gum and Glycerine, it is made as follows :—

In a glass measure, place sufficient clean picked gum arabic, to reach the line marking half an ounce, add cold distilled water to make one ounce. This should be left standing two days and stirred at intervals till completely dissolved. The solution is filtered by squeezing it two or three times through a fine linen cloth. One sixth part of a saturated solution of Boracic acid in Glycerine is now added, and the whole placed in a capsuled bottle containing a lump of camphor.

The formation of air bubbles by filtering and stirring is of no consequence, as they disappear completely after a few days.

BACILLUS TUBERCULOSIS.—Dr. Gibbes recommends staining the specimen with Methyline Blue instead of Chrysoidine, the acid being previously removed by frequent washings in distilled water.

This is a great improvement and if the Bacilli are mounted in Gum and Glycerine instead of Balsam, the reproduction of their appearance by means of photography is rendered comparitively easy.

BINOCULAR PHOTOGRAPHY. — Stereoscopic photographs may be produced by Stephenson's Binocular arrangement, Polycystinæ and Discoidal Diatoms form beautiful objects when seen in the Stereoscope.

The prisms should be placed as near the back of the objective as possible, and an ordinary stereoscopic camera substituted for that already mentioned.

Wenham's and most other arrangements are comparatively useless, as the light is not the same on both plates, and the plates themselves require to be placed at an inconvenient angle.

INDEX.

ERRATUM.

Pages 42—53, folio lines, *for* Hand &c., *read* Stand.

Printed by H. K. Lewis, 136 Gower Street London, W.C.